Energy Levels in Atoms and Molecules

W. G. Richards
University of Oxford

P. R. Scott
Royal Grammar School, Guildford

OXFORD NEW YORK TOKYO
OXFORD UNIVERSITY PRESS
1994

Oxford University Press, Walton Street, Oxford OX2 6DP

Oxford New York
Athens Auckland Bangkok Bombay
Calcutta Cape Town Dar es Salaam Delhi
Florence Hong Kong Istanbul Karachi
Kuala Lumpur Madras Madrid Melbourne
Mexico City Nairobi Paris Singapore
Taipei Tokyo Toronto

and associated companies in
Berlin Ibadan

Oxford is a trade mark of Oxford University Press

Published in the United States
by Oxford University Press Inc., New York

A catalogue record for this book is available from the British Library

Library of Congress Cataloging in Publication Data

Richards, W. G. (William Graham)
Energy levels in atoms and molecules / W.G. Richards, P.R. Scott.
(Oxford chemistry primers ; 26)
Includes index.
1. Energy levels (Quantum mechanics) 2. Nuclear magnetic resonance spectroscopy. I. Scott, P.R. (Peter Richard), 1950– II. Title. III. Series.
QD462.6.E53R53 1994 541.2'8—dc20 94-30693

ISBN 0 19 855804 X

Typeset using LaTeX
Printed in Great Britain by
The Alden Press, Oxford

Series Editor's Foreword

Oxford Chemistry Primers are designed to provide clear and concise introductions to a wide range of topics that may be encountered by chemistry students as they progress from the freshman stage through to graduation. The Physical Chemistry series will contain books easily recognized as relating to established fundamental core material that all chemists need to know, as well as books reflecting new directions and research trends in the subject, thereby anticipating (and perhaps encouraging) the evolution of modern undergraduate courses.

In this Physical Chemistry Primer Graham Richards and Peter Scott have produced a clear and easy-to-read account of how quantum mechanics helps us explain the diverse forms of energy associated with atoms and molecules. The Primer is designed to be of lasting value throughout any undergraduate chemistry course, starting from the earliest freshman days. This Primer will interest all students of chemistry (and their mentors).

Richard G. Compton
Physical Chemistry Laboratory, University of Oxford

Preface

At the most fundamental level, all of chemistry is a reflection of the ways in which electrons and nuclei interact with each other. The behaviour of electrons and nuclei are controlled by the rules of quantum mechanics — rules which are quite unlike those in the familiar world of classical mechanics, and which may at first seem quite complex.

In this book we try to show how quantum mechanics can explain the properties of atoms and molecules. A single electron orbiting a proton gives rise to the concept of orbitals, and more complex atoms are then described in terms of electron configurations. The overlap of valence orbitals causes atoms to come together to form molecules, and explains bond energies and molecular geometry. Quantum mechanics also determines how molecules rotate and vibrate, and how their nuclei behave when placed in strong magnetic fields. By describing the allowed states of simple atoms, molecules, and ions, the book therefore introduces the fundamental ideas of chemical bonding and energetics; and most importantly it provides the ideas needed to understand the various different branches of spectroscopy.

Our emphasis is unashamedly non-rigorous; the object is to provide a physical picture of how electrons and nuclei behave, and to show how reasonable approximations can shed light on the behaviour of quite complex systems. Many excellent texts consider a wider range of examples, and in greater mathematical detail; they will be the next step for those who wish to pursue these topics in greater depth. Our aim is to introduce an unfamiliar world, and to show that much of the behaviour of atoms and molecules can in fact be understood quite simply.

Oxford and Guildford
August 1994

W.G.R.
P.R.S.

Contents

1 Atoms

1.1 Waves and particles

Chemistry is concerned with the behaviour of electrons in atoms and molecules. When reactions occur, electrons are transferred or shared; a more favourable arrangement of electrons will result in an exothermic reaction, and therefore the formation of a more stable compound. We are used to talking of electron configurations, which describe the arrangement of electrons in shells or orbitals. From these configurations we make deductions about ionization energies, which help determine the stabilities of ionic compounds, and about bond energies, which control the reactions of covalent compounds.

However, we rarely stop to think where these ideas about electrons come from. It is not possible to measure directly in a simple calorimeter the energy required to remove an electron from a gas-phase atom, nor the energy evolved when a covalent bond is formed. Even ideas about the motions of electrons in orbitals or shells, though familiar from long use, are probably unsupported by any direct evidence. Classical models, such as the electron orbiting the nucleus as the earth orbits the sun, are full of contradictions. Why are only certain orbits allowed for electrons in atoms, while all orbits are possible for satellites? Is the hydrogen atom really planar, like the solar system, or spherical, as it is normally drawn?

In fact our ideas about electrons in atoms are soundly based, and derive from experiments performed at the beginning of the twentieth century. Only a hundred years ago some scientists believed that it made no sense to discuss chemistry in terms of the behaviour of atoms, as atoms could not be observed directly. The work of men like Planck, Bohr, and de Broglie led to a complete reappraisal of physics, which in turn has transformed chemistry from a sophisticated form of cookery into a discipline firmly based on theoretical foundations. Interpretation of atomic spectra gives detailed information about the motion of electrons in atoms. This in turn helps us to understand how and why atoms come together to form molecules, and also explains the structure of the Periodic Table.

To understand this information, we shall need to re-examine our understanding firstly of the nature of light, and then that of matter itself.

1.2 The photoelectric effect

What picture do we have of light? Is it a wave motion, rather like water waves, by which energy is transferred from one place to another without any matter being transferred? In which case, what exactly is the medium which oscillates, bearing in mind that light can travel through a vacuum? Alternatively, do we see light as a stream of particles? If so, what do the particles consist of; do they have mass, for example?

The question of whether light consists of waves or particles has been discussed since the seventeenth century, but it seemed to be resolved in the nineteenth century when it was shown that light could undergo interference and diffraction. These are well known properties of wave motions; their key feature is that a point illuminated by two sources of light can nevertheless be in darkness. This can be explained simply if we think of the two sources as producing waves which are exactly out of phase with each other at the point in question; it is much harder to see how two particles can come together to produce nothing. Measurements of diffraction experiments allowed the wavelengths of different forms of light to be determined.

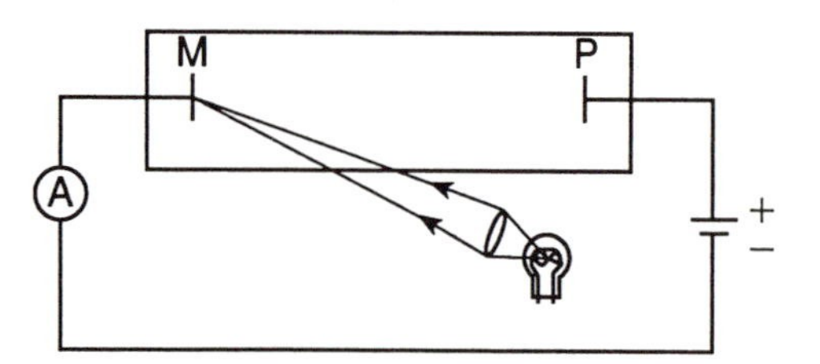

Fig. 1.1 Experimental arrangement for observing the photoelectric effect.

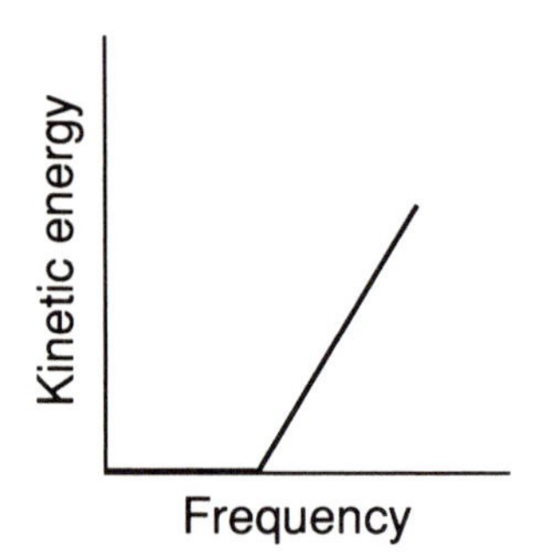

Fig. 1.2 Graph of maximum kinetic energy versus frequency in photoelectric effect experiment.

This picture was clouded, however, by the discovery of the photoelectric effect. A simple apparatus for this experiment is shown in Fig. 1.1. An evacuated chamber contains a piece of metal, M, and a plate, P. The plate is maintained at a slight positive potential, and the metal is kept at room temperature. With the lamp turned off, no electrons can pass from M to P, and so the ammeter, A, reads zero. However if ultraviolet light is allowed to fall on M, then electrons do pass from M to P, and the ammeter records a current. The important observation is that electrons are only emitted if the frequency of the light exceeds a certain critical value; this value varies from metal to metal. If the frequency falls below this critical value, then the current is zero, no matter what the intensity of the light is. By using monochromatic light, and by varying the potential of the plate P, the maximum energy with which the electrons are emitted can be measured as a function of the frequency of the light; the graph is shown in Fig. 1.2. The graph meets the horizonal axis at the critical frequency; below that no photoelectrons are produced. Above the critical frequency, the kinetic energy rises linearly.

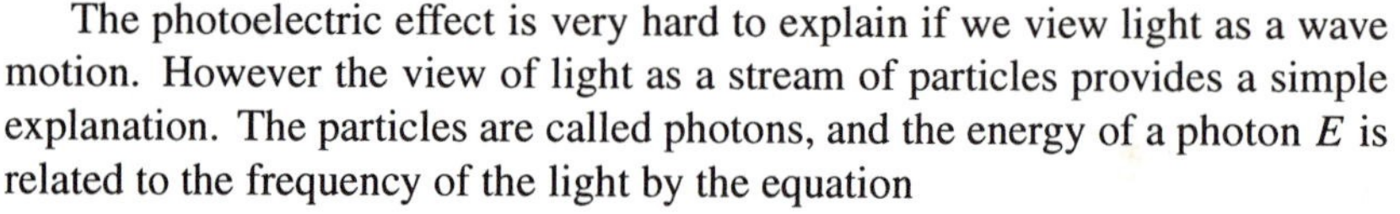

The photoelectric effect is very hard to explain if we view light as a wave motion. However the view of light as a stream of particles provides a simple explanation. The particles are called photons, and the energy of a photon E is related to the frequency of the light by the equation

$$E = h\nu,$$

where h is a constant called Planck's constant; its value is approximately 6.6×10^{-34} J s.

When a photon strikes the metal plate, its energy is transferred to an electron. Now energy is required to remove an electron from the metal, and this minimum energy is called the work function; the value varies from metal to metal. Any energy provided by the photon over and above the work function appears as the kinetic energy of the electron; the conservation of energy therefore gives us:

$$h\nu = \text{work function} + \text{kinetic energy}$$

This explains the appearance of the graph in Fig. 1.2. Electrons are only emitted from the metal when the energy of the photon exceeds the work function; this implies that a single quantum of energy must be transferred to a single electron, and that the energy of the light is localized, and not spread out across the whole wave front. If the frequency of the light is too low, then an electron does not acquire sufficient energy to escape from the metal; it cannot retain that energy, and wait for the arrival of a second photon.

1.3 Wave–particle duality

The key to the interpretation of the photoelectric effect is the equation:

$$E = h\nu.$$

This is a strange equation, as it relates a particle property, the energy of a photon, to a wave property, the frequency of the light (and hence the wavelength). The equation was first suggested from a consideration of an entirely different question, the nature of black-body radiation, where it helped to resolve a problem known as the ultraviolet catastrophe.

The photoelectric effect did not resolve the question of whether light should be thought of as waves or particles; rather it showed that the behaviour of light was not completely described by either simple model. It is now accepted that the behaviour of electromagnetic radiation is best described as a wave motion in some experiments, and as a stream of photons in others; this is known as wave–particle duality. This situation may seem to be rather unsatisfactory, although the problem arises partly because of our initial assumption that light ought to resemble models used to describe other phenomena in physics.

The balance was redressed somewhat when it was shown that electrons could be diffracted by the layers of atoms in regular crystals. Electrons had always been seen as particles; the Millikan oil-drop experiment had shown that the charge on small oil drops was always a simple multiple of what we now know to be the charge on the electron, and that the charge could not vary continuously. The electron diffraction experiment showed that a stream of electrons could behave as a wave, with the wavelength λ being related to the momentum p by the equation:

$$\lambda = h/p.$$

This equation, like $E = h\nu$, relates a particle property to a wave property.

Towards the end of the nineteenth century there were several attempts to provide an accurate description of the behaviour of electrons in atoms based on a view of the electron as a particle. Bohr produced a model of the hydrogen atom which gave some predictions which matched experimental data well, but only by making some unjustifiable assumptions about the behaviour of the electron as it circled the nucleus. Furthermore his model could not be extended to more complex atoms.

Eventually it was realized that the only way to understand atomic structure was to view the electron as a wave, not a particle. Schrödinger produced his famous wave equation, the solutions to which provide us with our modern description of atomic orbitals. We shall see later how the Schrödinger equation can be applied to the hydrogen atom; it provides a description of the behaviour of the electron in a hydrogen atom, and also predicts accurately how the atom will absorb and emit light. Spectroscopy, the study of the absorption and emission of light, gives us very precise data on atomic structure; this data can be understood if we think of electrons as waves, and light as particles. This is, of course, a complete reversal of the standard pictures of nineteenth-century physics.

1.4 Units and measurements

In this book we will be concerned with the various states in which atoms and molecules are found, and the energy differences between these states. Because much of the information about energy differences comes from spectroscopy, energies are often measured for convenience in terms of wavelengths and frequencies, rather than the more obvious Joules; the relationships between the various units that may be encountered are given below.

Wavelength

The energy differences between different electron configurations of an atom often correspond to the energies of photons in the visible and ultraviolet regions of the spectrum. Visible light has wavelengths approximately in the region 400–700 nm, with blue light corresponding to 400 nm, and red light 700 nm. To higher wavelengths we have infrared light, and to lower wavelengths ultraviolet.

Wavelengths are also measured in Angstroms; $1\ \text{Å} = 10^{-10}\ \text{m} = 0.1\ \text{nm}$. It therefore follows that light of wavelength 500 nm has wavelength 5000 Å.

Frequency

For all wave motions, wavelength, frequency, and velocity are related by the equation:

$$\text{wavelength} \times \text{frequency} = \text{velocity}$$

$$\lambda \times \nu = c.$$

The speed of light, c, is the same for light of all wavelengths, and so measurements of wavelengths can be converted readily to frequencies. For light of wavelength 500 nm,

$$\begin{aligned} \nu &= c/\lambda \\ &= 3 \times 10^8/500 \times 10^{-9} \\ &= 6 \times 10^{14}\ \text{Hz.} \end{aligned}$$

Wavenumbers

Another unit which is used in spectroscopy is the wavenumber, written cm^{-1}; it represents the number of waves which would occupy a length of 1 cm. Thus for light of wavelength 500 nm,

$$\begin{aligned} \lambda &= 5 \times 10^{-7}\ \text{m} \\ &= 5 \times 10^{-5}\ \text{cm} \\ \text{wavenumber} &= 1/(5 \times 10^{-5}) = 2 \times 10^4\ \text{cm}^{-1}. \end{aligned}$$

One advantage of using frequency and wavenumber, rather than wavelength, is that they are proportional to the energy of a photon, whereas the wavelength is inversely proportional. High energy therefore corresponds to high frequency, but low wavelength. Indeed, wavenumbers are often used directly as an energy unit.

Energy

The energy of an individual photon is given by:

$$E = h\nu,$$

where h is Planck's constant. So for light of wavelength 500 nm,

$$\begin{aligned} \nu &= 6 \times 10^{14}\ \text{Hz (from above)} \\ E &= 6.6 \times 10^{-34} \times 6 \times 10^{14}\ \text{J} \\ &= 4 \times 10^{-19}\ \text{J.} \end{aligned}$$

To compare this with thermochemical measurements, it is useful to calculate the energy per mole of photons:

$$\begin{aligned} E &= 4 \times 10^{-19} \times 6 \times 10^{23}\ \text{J mol}^{-1} \\ &= 2.4 \times 10^5\ \text{J mol}^{-1} \\ &= 240\ \text{kJ mol}^{-1}, \end{aligned}$$

which is comparable with the strengths of some covalent bonds.

The other unit commonly used is the electronvolt, the energy gained by an electron moving through a potential difference of 1 volt. The conversion is

$$1 \text{ eV} = 1.6 \times 10^{-19} \text{ J}$$

So for a photon of wavelength 500 nm,

$$\begin{aligned} E &= (4 \times 10^{-19})/(1.6 \times 10^{-19}) \\ &= 2.5 \text{ eV.} \end{aligned}$$

1.5 The electromagnetic spectrum

This book is concerned with widely varying energy differences; Fig. 1.3 shows the range of these energy differences in the various units described above, along with the corresponding parts of the electromagnetic spectrum.

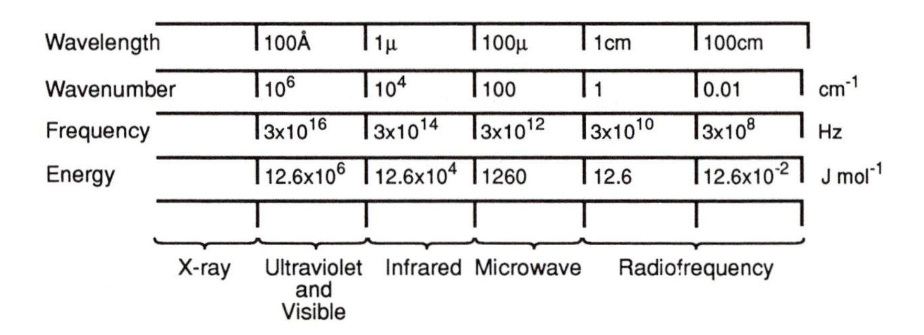

Fig. 1.3 The electromagnetic spectrum.

1.6 Energy levels of the hydrogen atom

The hydrogen atom is the simplest of all atoms, and is the starting point for our investigation of electronic structure. It is the simplest system because the electron experiences only the attraction of the proton at the centre of the atom; it is not repelled by other electrons in the atom.

Studies of the spectra of hydrogen atoms in the gas phase showed that these atoms could exist in a variety of different states, each with different energies. The most stable state, the one with lowest energy, is called the ground state; atoms in the ground state can undergo transitions to higher excited states, for example by absorbing UV light. In the excited states the electron is further from the nucleus than it is in the ground state.

Perhaps the most surprising feature of this is not that the electron can exist in a number of different states, but that only certain states with well-defined energies seem to be allowed. This is not true of planets orbiting a star for example, and it is not predicted by classical mechanics; it does, however, hold for all atoms. The energy of the atom is said to be quantized, that is it can only take certain values.

In the case of the hydrogen atom the energies of the allowed states are given by a simple formula,

$$E = -R/n^2$$

where R is a constant called the Rydberg constant, and n is any whole number from 1 to infinity. R therefore has the dimensions of energy, and has the value 13.46 eV; n is called a quantum number, and can be used to label the allowed states. For the ground state of the hydrogen atom n has its lowest possible value, 1, and so $E = -R$; for the first excited state n=2, and so $E = -R/4$; for the second excited state $E = -R/9$; and so on. The allowed states are shown in the energy level diagram in Fig. 1.4.

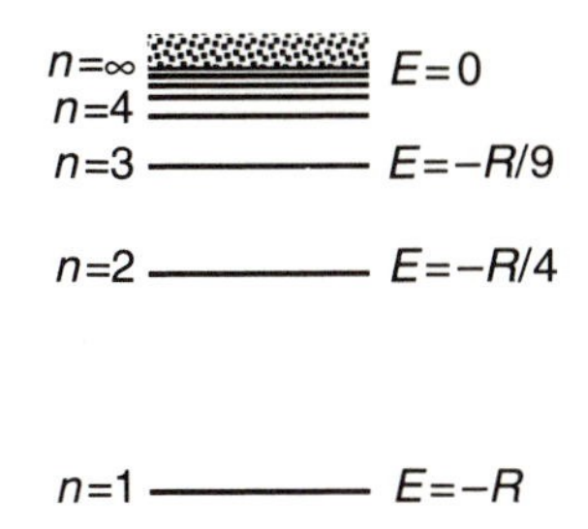

Fig. 1.4 Observed energy levels in the H atom.

When an atom in an excited state n_2 emits a photon and falls to a lower state n_1, the energy it loses is

$$\begin{aligned}\Delta E &= E_2 - E_1 = -R/n_2^2 + R/n_1^2 \\ &= R(1/n_1^2 - 1/n_2^2).\end{aligned}$$

This is equal to the energy of the photon, which we saw is given by $h\nu$, so

$$\nu = (R/h)(1/n_1^2 - 1/n_2^2),$$

which is exactly the result obtained from atomic spectroscopy.

As n gets larger and larger, the electron gets steadily further away from the proton, until eventually it is no longer bound at all. The quantum number n is now infinitely large, and so $E = 0$. The energy required to remove an electron from the ground state of a hydrogen atom to infinity is the ionization energy of the atom, and so

$$\begin{aligned}\text{IE} &= E_\infty - E_1 \\ &= 0 - (-R) \\ &= R \\ &= 13.46 \text{ eV}.\end{aligned}$$

The discovery that the allowed energy levels of the hydrogen atom were quantized, and that their energies were related to a quantum number by a simple formula, seemed to be an important breakthrough in understanding the electronic structures of atoms. In fact it turned out to be less helpful than expected. Firstly, the simple formula that described the hydrogen energy levels so accurately turned out to have no direct equivalent when other atoms were investigated. Secondly, it proved impossible to describe the behaviour of the electron in the hydrogen atom without ignoring the basic principles of classical physics; Newton's laws do not lead to quantized energy levels when an electron orbits a proton.

The breakthrough in understanding the hydrogen atom came when wave mechanics was applied to the problem; the electron is now viewed not as a particle, with a definite position and velocity, but rather as a wave, whose amplitude may be described. The next section introduces wave mechanics; at first it may seem complex, but the simple picture of an electron cloud that emerges is important, and can be understood without attention to the mathematical detail that precedes it.

1.7 Wave mechanics

The Schrödinger wave equation is the starting point for wave mechanics; it cannot be deduced from classical mechanics, nor can it be proved, other than by showing that its predictions agree with experiment. The electron is described by a wave function Ψ, which can be found by solving the equation

$$\frac{\mathrm{d}^2\Psi}{\mathrm{d}x^2} + \frac{\mathrm{d}^2\Psi}{\mathrm{d}y^2} + \frac{\mathrm{d}^2\Psi}{\mathrm{d}z^2} + \frac{8\pi^2 m}{h^2}(E - V)\Psi = 0.$$

The quantity Ψ represents the amplitude of the electron wave. Its interpretation is due mainly to Born, who by analogy with other wave equations, considered Ψ^2 as a probability. For a system containing one electron, $\Psi^2\,\mathrm{d}x$ is then the probability of finding the particle in the region of space $\mathrm{d}x$. (More correctly the probability is $\Psi^*\Psi\,\mathrm{d}x$, for Ψ may be a complex number; Ψ^* is the complex conjugate of Ψ). Alternatively we can visualize $\Psi^*\Psi\,\mathrm{d}x$ as being a measure of the density of matter in the region defined by $\mathrm{d}x$, with the particle having no discrete nature; this picture of particle clouds is a convenient, and generally accurate, way of thinking of wave functions.

The first three terms of the Schrödinger equation then represent double differentiation of the wave function, with respect to x, y, and z; E represents the total energy of the electron, and V its potential energy.

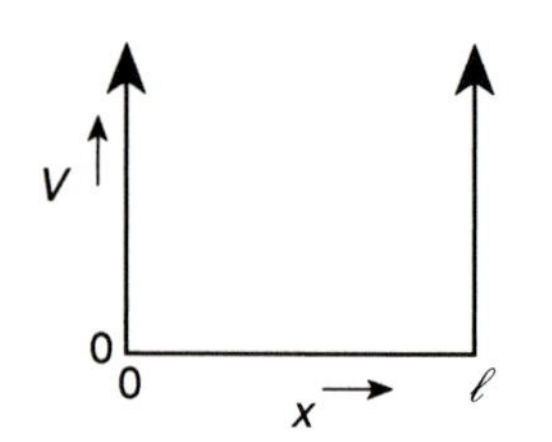

Fig. 1.5 Potential energy (V) of a particle in a one-dimensional box.

The Schrödinger equation looks complex, but its solutions are standing waves, whose forms are often simple. The behaviour of a particle in a one-dimensional box illustrates the implications of the equation. If a particle is confined to a box of length l, such that its potential energy V is zero inside the box, and infinite outside the box (Fig. 1.5), then the equation becomes:

$$\frac{\mathrm{d}^2\Psi}{\mathrm{d}x^2} + \frac{8\pi^2 m}{h^2}(E - V)\Psi = 0.$$

This equation has solutions of the type

$$\Psi_n = A \sin(n\pi x/l),$$

where n is a whole number. From the allowed wave functions Ψ_n we can calculate the corresponding allowed energies E_n; the wave functions, and the corresponding energy level diagram, are shown in Figs 1.6 and 1.7.

The particle in a box is not a model for a hydrogen atom, of course, but it shows how the wave model of the electron gives rise to the quantization

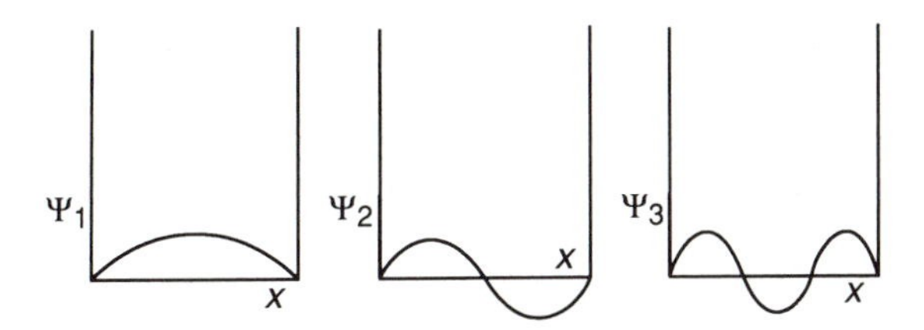

Fig. 1.6 Allowed wave functions for the particle in a one-dimensional box.

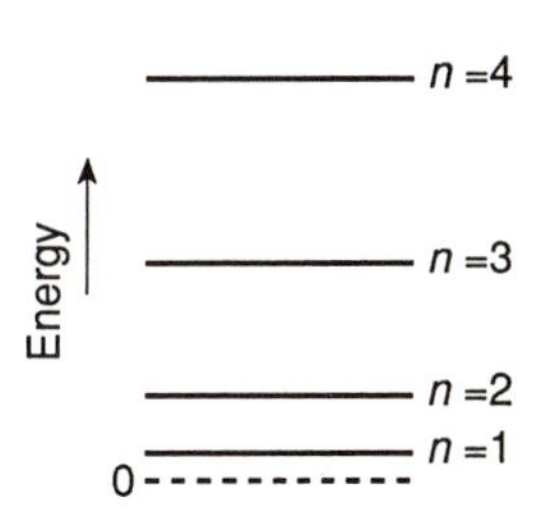

Fig. 1.7 Energy levels of a particle in a one-dimensional box.

of energy levels, through the requirement that standing waves are formed. A further point becomes clear when the problem is extended to a two-dimensional box; the resulting wave functions are the products of two terms, one described by a quantum number n_x, and the other by a second quantum number n_y:

$$\Psi = A \sin (n_x \pi x/l) \sin (n_y \pi y/l).$$

Adding a second dimension to the problem has created a second quantum number. It is, therefore, not surprising that when the Schrödinger equation is applied to the hydrogen atom, where the electron is free to move in three dimensions, the resulting wave functions contain three quantum numbers.

1.8 Quantum numbers

The Schrödinger equation for the electron in a hydrogen atom is

$$\frac{d^2\Psi}{dx^2} + \frac{d^2\Psi}{dy^2} + \frac{d^2\Psi}{dz^2} + \frac{8\pi^2 m}{h^2}(E - V)\Psi = 0.$$

In this case the potential V is the electrostatic nuclear-electron attraction $-e^2/4\pi\epsilon_0 r$.

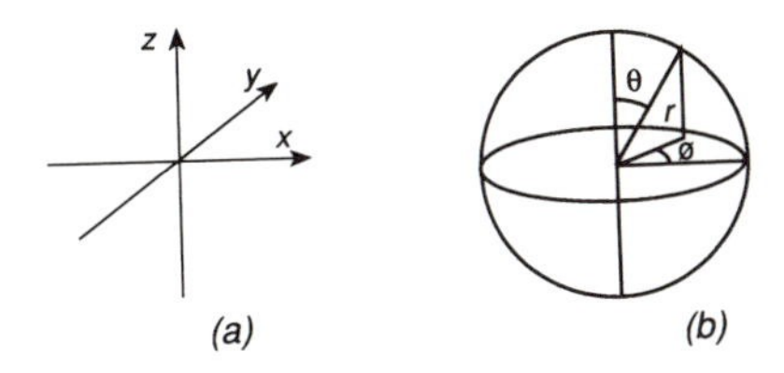

Fig. 1.8 (a) Cartesian (x, y, z) coordinates (b) Polar (r, θ, ϕ) coordinates.

The problem is more easily approached using spherical polar coordinates, r, θ, and ϕ, rather than the more usual Cartesian coordinates, x, y, and z. This alternative way of defining a point in space is illustrated in Fig. 1.8. Just as the particle in a two-dimensional box gave rise to a wave function which was a product of two terms, one in x and one in y, so the hydrogen wave functions may be expressed as a product of three terms, one in r, one in θ, and one in ϕ.

$$\Psi = R(r) \, . \, \Theta(\theta) \, . \, \Phi(\phi).$$

The actual steps in the solution of the Schrödinger equation are straightforward but lengthy, and are given in many books; we shall not rehearse them here. The mathematical functions $R(r)$, $\Theta(\theta)$, and $\Phi(\phi)$ are well known to mathematicians, and each contains, as expected, a quantum number; these are labelled n, l, and m_l respectively.

The quantum number n is called the principal quantum number, and it determines the energy of the atom. The wave equation predicts the relationship

$$E_n = -R/n^2,$$

which we met earlier in this chapter, and also correctly predicts the value of the Rydberg constant R.

The quantum number l is called the azimuthal quantum number, and it determines the value of the orbital angular momentum of the electron:

$$p = \sqrt{l(l+1)}\, h/2\pi$$

For historical reasons, electrons with $l = 0, 1, 2$, and 3 are labelled s, p, d, and f. The values which l can take are all whole numbers from 0 to $(n-1)$.

The third quantum number m_l is called the magnetic quantum number; it refers to the orientation of the orbital, and specifies the component of the angular momentum in a particular direction. It can take all whole number values from $+l$ to $-l$.

These three quantum numbers arise naturally from solving the Schrödinger equation for the hydrogen atom. As we shall see, they give us a detailed picture of how we may think of an electron in a hydrogen atom, and they also provide the basis for describing more complex atoms; indeed, the structure of the entire Periodic Table follows from these three quantum numbers.

Fig. 1.9 Radial variation of a 1s atomic orbital.

1.9 The hydrogen wave functions

We can now consider the detailed shapes of the wave functions produced by the Schrödinger equation; the simple picture we have is of an electron cloud, whose density at any point is measured by the value of Ψ^2.

1s orbital

For the ground state of the hydrogen atom, $n = 1$, and so l must be equal to 0, and m_l also equals 0. This we can describe as a 1s orbital, with the 1 representing the principle quantum number, and s representing the value of l. The wave function is

$$\Psi = N\mathrm{e}^{-r/a_0}$$

where N is a constant, r is the distance from the nucleus, and a_0 is a constant called the Bohr radius. The variation of the wave function with distance is shown in Fig. 1.9. In this case there is no variation with either ϕ or θ, and so the electron cloud is spherical; this is shown schematically in Fig. 1.10.

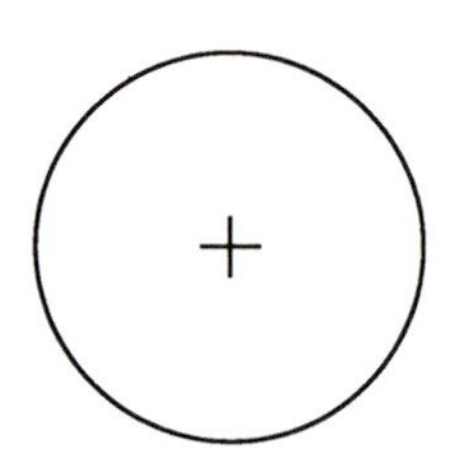

Fig. 1.10 Schematic representation of the three-dimensional variation of a 1s atomic wave function.

As the maximum value of l is $(n-1)$, it can only take the value 0, and so this orbital is the only one possible with $n = 1$.

2s orbital

For a 2s orbital, $n = 2$ and $l = 0$, and so $m_l = 0$. Just as for the 1s orbital there is no variation with either ϕ or θ, and the orbital is spherically symmetrical. The radial variation is shown in Fig. 1.11; although Ψ^2 is always positive, the sign of Ψ changes as the distance from the nucleus increases. This results in a node, that is a point where Ψ is zero. The main electron density in the 2s orbital is further away from the nucleus than in the 1s orbital, and so the energy of the orbital is higher.

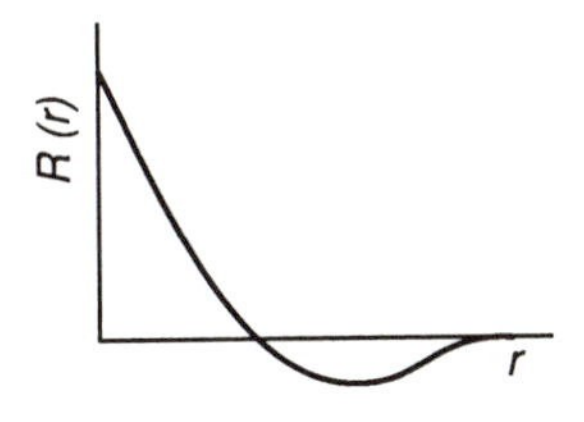

Fig. 1.11 Radial variation of an atomic 2s wave function.

2p orbital

If $n = 2$, then l can also take the value 1; now m_l can take three values, -1, 0, and $+1$. If we consider first the case where $m_l = 0$, then there is no dependence on ϕ, but the value of Ψ varies with both r and θ. The variation with r is shown in Fig. 1.12; here the value of Ψ is zero at the nucleus, a point which has consequences in magnetic resonance spectroscopy.

Ψ also varies with θ:

$$\Theta(\theta) = \cos\theta.$$

The value of $\Theta(\theta)$ therefore varies between -1 and $+1$, and for one direction has the value 0. The value of Ψ depends on the product of $R(r)$ and $\Theta(\theta)$, and therefore the wave function is positive for some values of θ, negative for others, and zero at right angles to the z-axis. As $\Phi(\phi) = 1$, there is no variation of the wave function with ϕ, and so the orbital is cylindrically symmetrical about the z-axis. This is summarized in the well known drawing of the 2p orbital in Fig. 1.13.

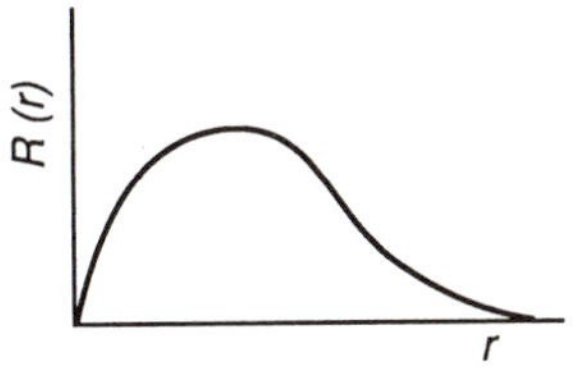

Fig. 1.12 Radial variation of an atomic 2p wave function.

m_l can also take the values -1 and $+1$; this corresponds to the orbitals having the same size and shape as the orbital in Fig. 1.13; only the orientation of the orbital changes. This is often represented by two further 2p orbitals, with their lobes directed along the x and y axes. (Strictly speaking this is not exactly correct, as the combinations $(p_x + ip_y)$ and $(p_x - ip_y)$ are needed to give the correct values of m_l; nevertheless the orbitals in the diagram act as useful chemical models.)

Although the wave functions for 2s and 2p orbitals have quite different shapes, the orbitals do have energies which are exactly identical, that is they are degenerate.

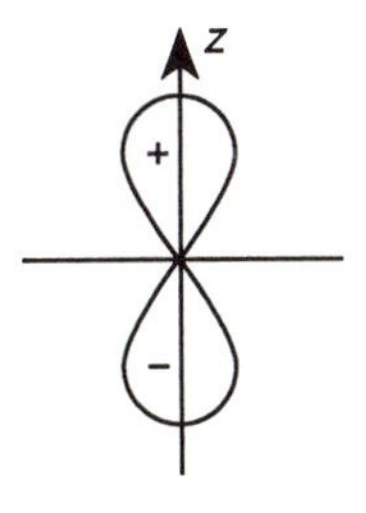

Fig. 1.13 Representation of the three-dimensional variation of an atomic 2p wave function.

3s, 3p, and 3d orbitals

If $n = 3$, then l may take the values 0, 1, and 2, and so we have 3s, 3p, and 3d orbitals, which are all degenerate with each other.

The 3s orbital is again spherically symmetrical; its electron density has a higher average distance from the nucleus than in the 1s and 2s cases, so its energy is correspondingly higher. Its radial function is more complex, now having two nodes. The three 3p orbitals have the same angular functions as the 2p orbitals, but once again the average distance of the electron from the nucleus is greater; the radial function now has one node.

There are five 3d orbitals, corresponding to values of m_l $-2, -1, 0, +1,$ and $+2$. Their angular distributions are complex, but their radial distribution is simpler, and has no nodes. The radial distributions of 3s, 3p, and 3d orbitals are shown in Fig. 1.14.

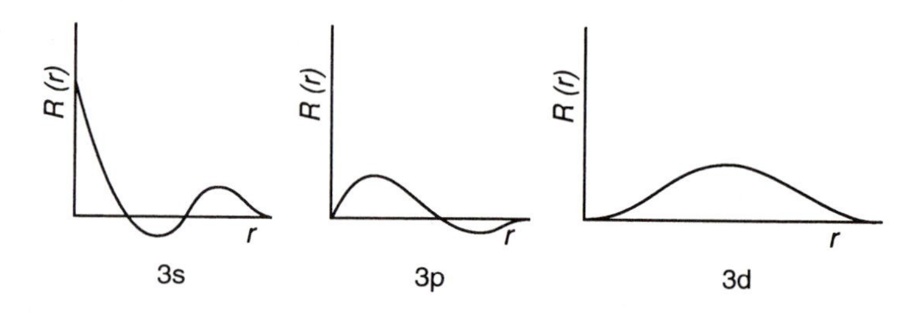

Fig. 1.14 Radial distributions of 3s, 3p, and 3d orbitals.

Orbital energies

Applying the Schrödinger wave equation to the hydrogen atom has led us to a series of allowed solutions, described by three quantum numbers n, l, and m_l. n can be any positive whole number; l can be any positive whole number up to $(n-1)$, and m_l can be any whole number between $+l$ and $-l$. The energies of these allowed solutions are given by

$$E_n = -R/n^2.$$

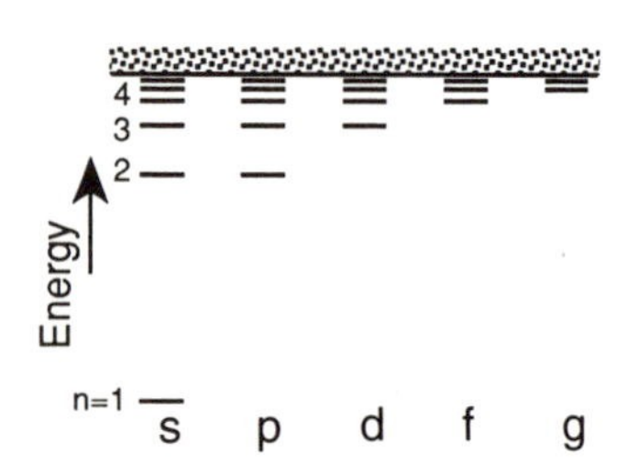

Fig. 1.15 Schematic energy level diagram for the hydrogen atom.

We can now draw a new version of the hydrogen energy-level diagram (Fig. 1.15). In many ways it resembles that in Fig. 1.4 which was obtained from spectroscopic measurements, but it also shows that many of the levels in the earlier diagram are in fact degenerate, that is consist of two or more states with the same energy. It is of course encouraging that our theoretical model agrees with experimental observations; more importantly, we shall see that the model is capable of being extended to deal with more complex situations, such as the introduction of further electrons, or the application of magnetic fields. In these situations, the degeneracy of orbitals with the same value of n disappears.

1.10 Electron spin

We have seen that applying the Schrödinger equation to a one-dimensional problem gives rise to one quantum number, and to a three-dimensional problem leads to three quantum numbers.

However, our description of the hydrogen atom as a three-dimensional problem is not quite complete, as it takes no account of the effects of relativity. In relativity, time is added as a fourth dimension, and when Dirac modified the Schrödinger equation to take account of relativity, he found that his equations required the existence of a fourth quantum number.

Historically the need for a fourth quantum number became clear from experiment, rather than theory. Uhlenbeck and Goudsmit introduced the electron spin angular momentum quantum number s, which has the value $\frac{1}{2}$; the angular momentum was pictured as arising from the electron spinning on its own axis (Fig. 1.16). The magnitude of the spin angular momentum is then $\sqrt{\frac{1}{2}(\frac{1}{2}+1)}\ h/2\pi$ (compare this with the equation for the orbital angular momentum in Section 1.8); it can be oriented in two possible directions, corresponding to the fourth quantum number m_s taking the values $+\frac{1}{2}$ and $-\frac{1}{2}$.

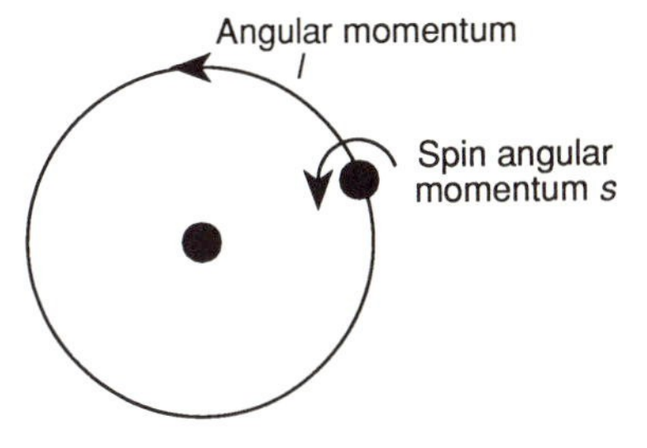

Fig. 1.16 Spin and orbital angular momentum.

The quantum number m_s was introduced to explain the observation that lines in the spectrum of the alkali metals appeared in closely spaced pairs; an electron with orbital quantum number l would then have two possible values of the total angular momentum $(l+\frac{1}{2})$ and $(l-\frac{1}{2})$, giving rise to two states of slightly different energies. The idea of electron spin was further supported by the Stern–Gerlach experiment, in which a beam of alkali metal atoms, with no orbital angular momentum, was split into two when passed through an inhomogeneous magnetic field; one beam contained atoms with $m_s = \frac{1}{2}$, and the other with $m_s = -\frac{1}{2}$.

1.11 The Pauli principle

It might be thought that the electronic structures of all atoms could be calculated in the same way that we used for the hydrogen atom in Section 1.8; writing the Schrödinger equation, separating the problem into radial and angular parts, and solving the resulting equations. Unfortunately this is not the case, as the terms describing the repulsion between electrons prevent the equation from being solvable in this way. The Schrödinger equation can only be solved analytically when there is one electron orbiting the nucleus.

However, we can use the results that we obtained for the hydrogen atom earlier in the 'orbital approximation'; the wave function of the atom is written as the product of orbitals (which are just mathematical functions describing the behaviour of each electron individually). Each electron then has its own value of n, l, m_l, and m_s.

We now need one further idea if we are to make sense of the electronic structures of complex atoms. The Pauli principle restricts the possible values that the quantum numbers of individual electrons can have. In its common, though less fundamental form, it states that:

no two electrons in an atom may have all four quantum numbers the same.

Every pair of electrons in an atom must therefore differ in at least one quantum number.

This form of the Pauli principle is a consequence of a more basic postulate about the wave functions of electrons. The square of the wave function Ψ^2 is a measure of the probability of finding electrons at a particular point in space.

If we label the electrons 1, 2, 3, etc, and then exchange two electrons, the electron density will not change, as all electrons are identical. It therefore follows that Ψ^2 must be unchanged if two electrons are exchanged. There are two possibilities which would allow this; one is that Ψ should remain unchanged on exchange of electrons, the other is that it should simply change sign ($\Psi \times \Psi = -\Psi \times -\Psi = \Psi^2$). It is found universally in nature that particles of spin $\frac{1}{2}$, such as electrons, behave such that an exchange of identical particles causes the wave function to change sign. This is called antisymmetric behaviour. The more fundamental form of the Pauli principle is therefore:

wave functions must be antisymmetric with respect to particle exchange.

At first sight this statement may seem very obscure, and unrelated to the earlier statement of the Pauli principle; in fact its consequences are enormously important, as it prevents all the electrons in an atom from occupying the 1s orbital. Without the Pauli principle, the rich variety in the chemistry of the elements would not exist.

Consider the He atom, which has two electrons. They both occupy the 1s orbital, that is they both have $n = 1, l = 0$, and $m_l = 0$. If we denote the two possible values of m_s, $+\frac{1}{2}$, and $-\frac{1}{2}$, as α and β, and label the electrons 1 and 2, then it might be thought that the wave function could be written:

$$\Psi = 1s^{\alpha}(1)\ 1s^{\beta}(2).$$

However this is not satisfactory, as changing the labels 1 and 2 changes the wave function. Instead we must use the linear combination:

$$\Psi = 1s^{\alpha}(1)\ 1s^{\beta}(2) - 1s^{\alpha}(2)\ 1s^{\beta}(1).$$

If we exchange the electrons, 1 and 2 are interchanged, the first term becomes the second and vice versa, and so Ψ becomes $-\Psi$. Ψ^2, the electron density, is of course unchanged.

It is always possible to produce such a linear combination which satisfies the Pauli principle, provided that there is at least one quantum number different between the two electrons. However if two electrons have four identical quantum numbers, the two terms become identical, and the wave function becomes zero. We have considered here only the ground state of the He atom, but the same procedure could be used with any pair of electrons in any atom, and the conclusion would be the same; if we are to form an antisymmetric wave function, then no two electrons can have identical quantum numbers.

1.12 The Periodic Table

We are now in a position to consider the electronic structures of polyelectronic atoms. We may begin by starting with a H atom, in which there is one electron in the 1s orbital, and proceed to the He atom by increasing the nuclear charge by one, and by adding one more electron (we can ignore the fact that this would be almost impossible to achieve in practice). The new electron can also occupy the 1s orbital, provided it has the opposite spin to the first electron; the configuration of He is therefore $1s^2$.

Table 1.1 Electron configurations

H	1s	C	$1s^2\ 2s^2\ 2p^2$
He	$1s^2$	N	$1s^2\ 2s^2\ 2p^3$
Li	$1s^2\ 2s$	O	$1s^2\ 2s^2\ 2p^4$
Be	$1s^2\ 2s^2$	F	$1s^2\ 2s^2\ 2p^5$
B	$1s^2\ 2s^2\ 2p$	Ne	$1s^2\ 2s^2\ 2p^6$

If we now add a further electron, and increase the nuclear charge again, we have a Li atom. Now the 1s orbital is full, and cannot hold another electron without contravening the Pauli principle, and so the electron goes into an orbital with $n = 2$. In the H atom the 2s and 2p orbitals had identical energies, but as we shall discuss in Section 1.14, in other atoms the 2s orbital is slightly more stable than the 2p; Li therefore has the configuration $1s^2\ 2s$. We can go on building up a table of electron configurations in this way, which is called the *aufbau* principle; the results for atoms up to Ne are given in Table 1.1.

We can now predict the electron configurations of all the remaining elements in the Periodic Table, provided that we know the order in which the orbitals are filled. In hydrogen, all orbitals with the same value of the principal quantum number n are degenerate, but in other atoms this degeneracy is lifted; orbitals with lower values of l are of lower energy, and so 3s orbitals are filled before 3p, and 3p before 3d. The splitting between orbitals of different l values can be greater than the energy gap between orbitals with different values of n; thus the 4s orbital is filled before the 3d orbital, and the 5s before the 4f.

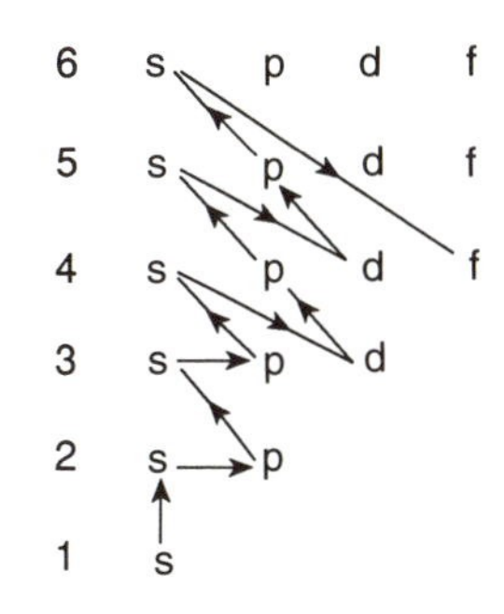

Fig. 1.17 Energy sequence of atomic orbitals.

The order in which the orbitals are filled is usually:

$$1s < 2s < 2p < 3s < 3p < 4s < 3d < 4p < 5s < 4d < 5p < 6s < 4f$$

as shown in Fig. 1.17. This order is predicted if the Schrödinger equation is solved by numerical methods, and ties in exactly with the familiar structure of the Periodic Table (inside back cover). In the Periodic Table the elements occur in blocks of 2, 6, 10, or 14; we can now recognize these in the light of the *aufbau* principle. A block of 2 elements corresponds to an s orbital being filled; here $l = 0$, so $m_l = 0$, and $m_s = +\frac{1}{2}$ or $-\frac{1}{2}$, so there are only two different sets of quantum numbers possible. When a p orbital is being filled, we see a block of six elements; m_l can take three values, -1, 0 or $+1$, and m_s two values, $+\frac{1}{2}$ or $-\frac{1}{2}$, and so six different sets of quantum numbers are possible. Similarly, filling a d orbital gives rise to a block of 10 elements, and an f orbital to a block of 14. The blocks occur in the order given in Fig. 1.17.

The *aufbau* principle does not take explicit account of variations in the electron repulsion between different orbitals; these can be significant if two orbitals have similar energies. This accounts for small deviations from the *aufbau* predictions in the first transition series (Table 1.2); in this series the 4s and 3d orbitals are very close in energy.

Table 1.2 Electron configurations of the first transition series

	Ar core + 4s	+ 3d
Sc	2	1
Ti	2	2
V	2	3
Cr	1	5
Mn	2	5
Fe	2	6
Co	2	7
Ni	2	8
Cu	1	10
Zn	2	10

Much of the basic physics and chemistry of the elements can be rationalized in terms of the electron configurations of atoms, but it is not the task of this book to pursue this story. What is important here is to see how the detailed shape and structure of the Periodic Table follows just from the quantum numbers which arose in solving the Schrödinger equation for the H atom.

1.13 Hund's rule

The basis of the *aufbau* principle is the idea that electrons will occupy the orbitals of lowest energy possible, subject to the requirements of the Pauli principle. Although this predicts many electron configurations satisfactorily, it leaves unresolved the question of how the electrons are arranged when a p, d, or f shell is partially occupied. An example of this would be the N atom, whose configuration is $1s^2\ 2s^2\ 2p^3$.

There are three different 2p orbitals, all with the same radial distribution, but with different orientations in space; they therefore all have the same energy. The best distribution of three electrons among these orbitals is given by Hund's rule, which states that the electrons arrange themselves so as to have as many spins parallel as the Pauli principle will allow. So for a N atom, the quantum numbers of the seven electrons are:

n	l	m_l	m_s	
1	0	0	$+\frac{1}{2}$	
1	0	0	$-\frac{1}{2}$	($1s^2$)
2	0	0	$+\frac{1}{2}$	
2	0	0	$-\frac{1}{2}$	($2s^2$)
2	1	-1	$+\frac{1}{2}$	
2	1	0	$+\frac{1}{2}$	
2	1	$+1$	$+\frac{1}{2}$	($2p^3$).

In order to have parallel spins, the electrons must occupy different orbitals, and this reduces the repulsion between the electrons. In the same way, a carbon atom has two of its 2p orbitals singly occupied, and one orbital empty, and an oxygen atom has one 2p orbital doubly occupied, and the other two orbitals singly occupied.

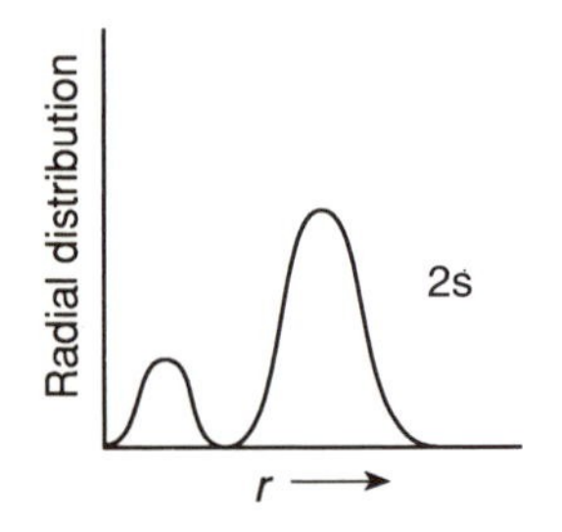

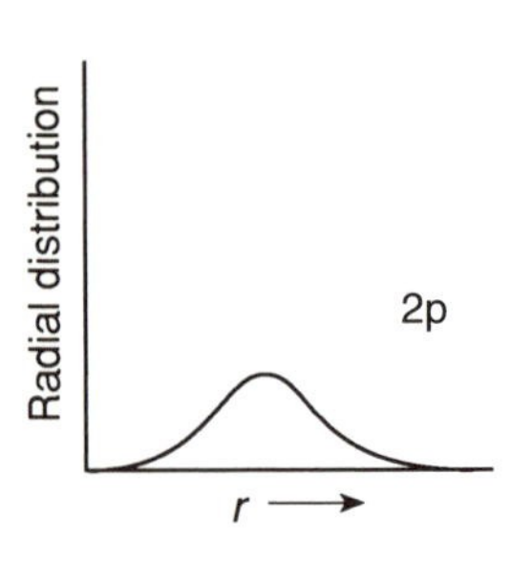

Fig. 1.18 Radial distributions of 2s and 2p atomic orbitals.

1.14 2s and 2p orbitals

One feature of the electronic structure of polyelectronic atoms is that 2s electrons are of slightly lower energy than 2p electrons, while earlier we saw that in the H atom they are degenerate. This unexpected difference can be explained by the radial distributions R of the two orbitals, which are shown in Fig. 1.18. The figure also shows the radial charge density $4\pi r^2 R^2$, which measures the total electron density at a distance r.

For the 1s orbital, the main electron density is close to the nucleus; for the 2s orbital the main electron density is further from the nucleus, but there is also a significant electron density very close to the nucleus. The 2p orbital has its maximum electron density slightly closer to the nucleus than the 2s orbital, but does not have such a high density very close to the nucleus. In the H atom, the result of this is that the 2s and 2p orbitals are degenerate.

The situation changes however if we consider an example such as Li, in which the configuration $1s^2$ 2s is now more stable than $1s^2$ 2p. When the outer electron is in the 2p orbital, almost all its density lies further from the nucleus than the 1s electrons, and so it experiences an effective positive charge of +1, made up of +3 from the nucleus, and −2 from the 1s electrons. However if the outer electron is in a 2s orbital, most of its electron density lies beyond the 1s electrons, thus experiencing an effective nuclear charge of +1, but a small proportion of its electron density now lies closer to the nucleus than the 1s electrons, and it experiences an effective charge of +3. The presence of the 1s electrons therefore favours the 2s orbital over the 2p; the 2s orbital is said to penetrate the 1s orbital, thereby increasing its stability.

1.15 4s and 3d orbitals

Penetration of inner electron shells also explains another strange feature of the order in which orbitals are filled, namely the relative energies of the 4s and 3d orbitals.

In the H atom, the 4s orbital has a higher energy than the 3d, as its principle quantum number *n* is higher. But the K atom has 19 electrons, 18 of which fill the orbitals 1s to 3p, and 1 further electron in the 4s orbital. The state in which the last electron occupies a 3d orbital is an excited state. We can understand this if we consider the radial distribution functions of the 4s and 3d orbitals, which are shown in Fig. 1.19.

Most of the electron density of the 3d orbital is closer to the nucleus than the main concentration of the 4s electron density; that is why the 4s orbital is less stable than the 3d orbital in H. But in K, the 3d electrons are well shielded from the nucleus, and experience an effective charge of about +1; in contrast the 4s orbital is able to penetrate the inner shells, because a small proportion of its density lies much closer to the nucleus. It therefore experiences an effective nuclear charge greater than +1, and this makes the 4s orbital more stable than the 3d orbital in this case.

In fact the 4s and 3d orbitals are very close to each other in energy, and the order is not always as it is in the K atom. Increasing the nuclear charge favours the 3d orbital proportionally more than the 4s orbital, and so although K has the ground state [Ar] 4s, the Sc^{2+} ion, which has the same number of electrons, has the ground state [Ar] 3d. In transition-metal atoms the 4s orbital is usually more stable than the 3d orbital, while in positive ions the reverse is true.

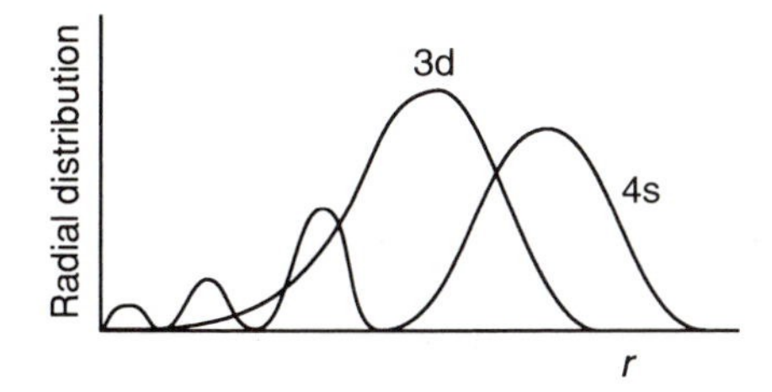

Fig. 1.19 Radial distributions of 3d and 4s atomic orbitals.

1.16 Ionization energies

So far we have been concerned primarily with the electronic ground states of atoms; however electrons can be excited to less stable orbitals, just as we saw in the H atom, and this gives rise to excited states.

In the H atom there is a series of excited states, whose energies follow a simple formula:

$$E_n = -R/n^2.$$

As the electron moves further from the nucleus, it becomes less tightly bound, and the energies of the excited states converge on a limit which represents the ionization energy of the atom.

There is no simple formula which will predict the energies of all excited states of polyelectronic atoms, but qualitatively the situation is similar; an electron can be excited to progressively higher orbitals, and their energies converge on a series limit representing an ionization energy. In many cases the energies of such excited states have been represented by the formula:

$$E_n = -R/(n - \delta)^2,$$

where δ is an empirical correction called the quantum defect. In polyelectronic atoms the various types of electron (s, p, d, etc.) penetrate the inner shells to different extents, and hence experience different effective nuclear charges. Table 1.3 gives the values of δ for the states of Na arising from the configuration $1s^2\ 2s^2\ 2p^6\ nl$. The quantum defect is seen to decrease in the order s > p > d, which is consistent with the analysis in the previous two sections of the relative penetrations of these orbitals. In very highly excited states, the quantum defect is negligible, and the inner electrons can be considered as point charges at the nucleus. In these circumstances the excited states resemble those of the H atom closely; these so-called Rydberg states are also observed in small molecules.

Table 1.3 Values of δ, the quantum defect, for some states of Na

$l =$	0	1	2	3	4
	s	p	d	f	g
$n = 10$	1.35	0.86	—	0.00	0.00
7	1.35	0.86	0.01	0.00	0.00
5	1.35	0.86	0.01	0.00	0.00
4	1.36	0.87	0.01	0.00	—
3	1.37	0.88	0.01	—	—

Observation of series limits by visible and U.V. spectroscopy, either directly or by extrapolation, represents the most accurate method of determining ionization energies; observations on positively charged ions allow second and higher ionization energies to be measured. Knowledge of ionization energies is essential for all calculations based on the ionic model of the stabilities of compounds in the solid state and in aqueous solution.

1.17 The energy levels of atoms

So far we have seen what the orbitals of a hydrogen atom are, and we have used these as a basis for the description of the structures of more complex atoms. We can describe the state of an atom by its electron configuration; but this, it turns out, is not yet a sufficient description.

The electron configuration describes the behaviour of each electron in the atom in terms of quantum numbers, which specify its orbital and spin angular momenta. We must now see how the angular momenta of the electrons combine to give a total angular momentum for the whole atom. This is important as it affects the total energy of the atom; one electron configuration may well give rise to a number of different states, each with different energies. The interaction between different angular momenta is sometimes called coupling.

Our final description of the state of an atom will therefore consist not only of its electron configuration, but also of quantum numbers describing its various forms of angular momentum. These are important not only because they affect the energy of the atom, but also because they determine to what other states the atom may jump by absorption or emission of light; that is they provide the vocabulary in which spectroscopic selection rules are expressed.

1.18 Coupling of angular momenta

Angular momentum can be thought of as a vector quantity, that is it can be represented by an arrow, whose length represents its magnitude, and whose orientation represents its direction. For angular momentum, the arrow points at right angles to the plane of rotation, as shown in Fig. 1.20.

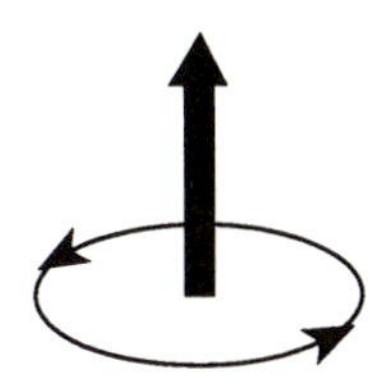

Fig. 1.20 Vector representation of angular momentum.

In classical mechanics, vectors (such as forces or velocities) are added by forming a triangle, as in Fig. 1.21; the resultant then depends on magnitudes and directions of the original vectors. Thus addition of one force to another may result in the force being increased or decreased, and its direction may stay the same, or be changed.

In quantum mechanics the situation is rather simpler, and no trigonometry is needed to calculate resultants. Just as the individual angular momenta are quantized, so are the resultants. If two angular momenta are coupled together, then the maximum value of the resultant is represented by the sum of the quantum numbers, and the minimum value by the difference. Intermediate values are allowed too, provided they differ from the extremes by a whole number. Thus when an angular momentum whose quantum number is 1 couples with one whose quantum number is $\frac{1}{2}$, the resultant may be $1\frac{1}{2}$ or $\frac{1}{2}$; when one of quantum number 2 couples with one of 1, the resultant may be 3, 2, or 1. The examples contained in the rest of this chapter will illustrate the application of these rules.

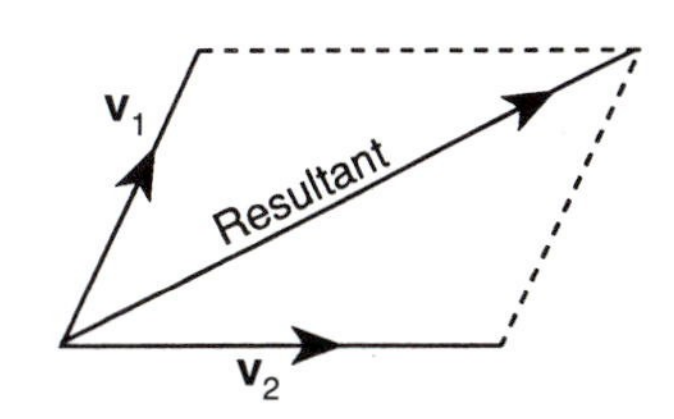

Fig. 1.21 The resultant of two vectors v_1 and v_2.

1.19 Russell–Saunders coupling

Energy differences between the various states of an atom can arise because of the differences in electrostatic repulsions between electrons, or because of the magnetic interactions between the intrinsic magnetic moments of the electrons and the fields created by their orbital motions. In the Russell–Saunders coupling scheme, with which we shall be primarily interested, the magnetic effects are treated as being negligible to a first approximation; this turns out to be a very good description of light atoms. In heavier atoms the approximation is no longer realistic, and we will consider in Section 1.25 how such atoms can be described.

For Russell–Saunders coupling the procedure for finding the states arising from a particular electron configuration is given below; we will then consider how it applies to some real examples.

1. The individual spin angular momenta of the electrons s (each of which has the value $\frac{1}{2}$) combine to give a resultant S:

$$\Sigma\, s = S.$$

   Thus two electrons, each with $s = \frac{1}{2}$, may combine to give $S = 0$ or 1; s describes only the magnitude of the spin, not its direction, so the spins may reinforce each other or cancel out.

2. The individual orbital angular momenta of the electrons l combine to give a resultant orbital angular momentum L:

$$\Sigma l = L.$$

   Remember that s electrons have $l = 0$, p electrons $l = 1$, and d electrons $l = 2$; two p electrons could therefore produce a resultant $L = 2, 1,$ or 0.

3. Now L and S couple together to give a total resultant angular momentum J. As we expect, J is also quantized, and can take the values $(L + S)$, $(L + S - 1), \ldots |L - S|$. Figure 1.22 illustrates how the vectors combine.

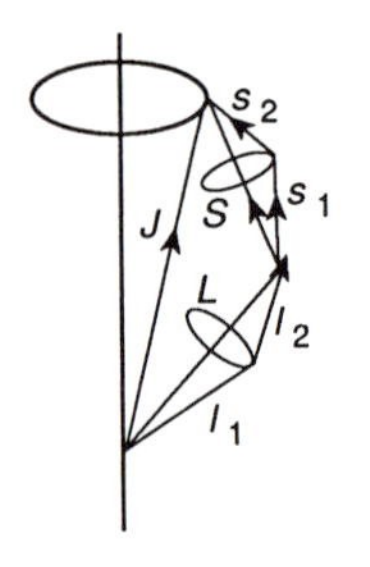

Fig. 1.22 The Russell–Saunders coupling scheme.

1.20 Term symbols

We have seen that the coupling of angular momenta can lead to the existence of several different states of an atom, all derived from a single electron configuration. For historical reasons these different states, each with their own energy, are sometimes referred to as 'terms', and they are described by 'term symbols', which contain information about the values of the quantum numbers L, S, and J.

Again for historical reasons, term symbols are written in a curious way, which appears strange at first, but which soon becomes familiar. Firstly, the value of the quantum number L is not recorded as a number, but as a letter, just as individual electrons with $l = 0$, 1, and 2 are described as s, p, and d. Thus a term with $L = 0$ is called S, with $L = 1$, P, with $L = 2$, D, and so on. Secondly, the value of S is not recorded directly, but rather the value of $(2S + 1)$. Thus a state with $S = 0$ has $(2S + 1) = 1$, and the state is called a singlet. If $S = \frac{1}{2}$, then $(2S + 1) = 2$, and the state is a doublet. $S = 1$ gives rise to a triplet, and so on. The whole term symbol is now written:

$$^{2S+1}L_J$$

Typical term symbols would therefore be:

$$^1\mathrm{S}_0 \quad ^2\mathrm{S}_{\frac{1}{2}} \quad ^3\mathrm{P}_2$$

which are read as 'singlet S nought', 'doublet S a half', and 'triplet P two' respectively.

1.21 Closed electron shells

Before we consider the term symbols for some real atoms, there is one important simplifying feature to be identified. Whenever we have a complete shell or sub-shell of electrons, such as s^2, p^6, or d^{10}, then these electrons have resultant spin and orbital angular momenta of zero.

If we consider a pair of 1s electrons in He, then we know from the Pauli principle that the two electrons cannot have four identical quantum numbers. They have the same values of n, l, and m_l, and so one must have $m_s = \frac{1}{2}$, and the other $m_s = -\frac{1}{2}$. The resultant total spin angular momentum is therefore zero. Similarly for a $2p^6$ configuration, as in Ne, for every electron with value $m_l = 1$, there is another electron with $m_l = -1$, and so the total value of the orbital angular momentum of the whole atom is again zero.

We can therefore write down the term symbols for the ground states of all the noble gases, which have only complete electron shells:

He	$1s^2$
Ne	$1s^2\ 2s^2\ 2p^6$
Ar	$1s^2\ 2s^2\ 2p^6\ 3s^2\ 3p^6$.

Each has a term symbol $^1\mathrm{S}_0$, corresponding to no spin or orbital angular momentum.

Furthermore even with atoms which contain many electrons, it is rarely necessary to consider more than a few of the electrons in calculating term symbols. All electrons in closed shells contribute nothing to L, S, or J, and may be ignored. Thus the Group 1 metals (such as Na, $1s^2\ 2s^2\ 2p^6\ 3s^1$) all reduce to one electron problems, provided that we do not excite an electron out of the closed shell core. Similarly in the transition metals it is only the valence electrons, which are usually d electrons, which influence the term symbols.

1.22 Some simple atoms

Hydrogen

The ground state of the hydrogen atom has one electron in a 1s orbital; as there is only one electron, there is no difference between s and S, or between l and L. Therefore

$$L = 0 \text{ and } S = \tfrac{1}{2},$$

with the result that $J = \frac{1}{2}$. The term symbol for the atom is therefore $^2S_{\frac{1}{2}}$. The situation is exactly the same for excited states where the electron is in an s orbital, such as 2s or 3s. Again $L = 0$, $S = \frac{1}{2}$, and $J = \frac{1}{2}$; the term symbol is $^2S_{\frac{1}{2}}$. A slightly different situation arises if the electron is excited to a p orbital, such as 2p. Now

$$L = 1 \text{ and } S = \tfrac{1}{2},$$

with the result that total angular momentum J can take the values $\frac{1}{2}$ or $\frac{3}{2}$. There are therefore two states with term symbols $^2\mathrm{S}_{\frac{3}{2}}$ and $^2\mathrm{S}_{\frac{1}{2}}$. In the absence of magnetic effects these two states would have identical energies, and be degenerate; in practice, the spin and orbital angular momenta do interact slightly. In the case of the 2p electron in a hydrogen atom, the splitting between the $^2\mathrm{S}_{\frac{3}{2}}$ and $^2\mathrm{S}_{\frac{1}{2}}$ states is only 0.1 cm^{-1}; this compares with a difference of 83000 cm^{-1} between the 1s and 2p configurations.

If we consider the term symbols arising from a hydrogen atom having an electron in a d orbital, such as 3d, then

$$L = 2 \text{ and } S = \tfrac{1}{2},$$

with the result that J can be $\frac{5}{2}$ or $\frac{3}{2}$. Again there are two terms, $^2\mathrm{D}_{\frac{5}{2}}$ and $^2\mathrm{D}_{\frac{3}{2}}$, which are very close in energy. Remember that the lowest allowed value of J is $|L - S|$, so that there is no state with $J = \frac{1}{2}$.

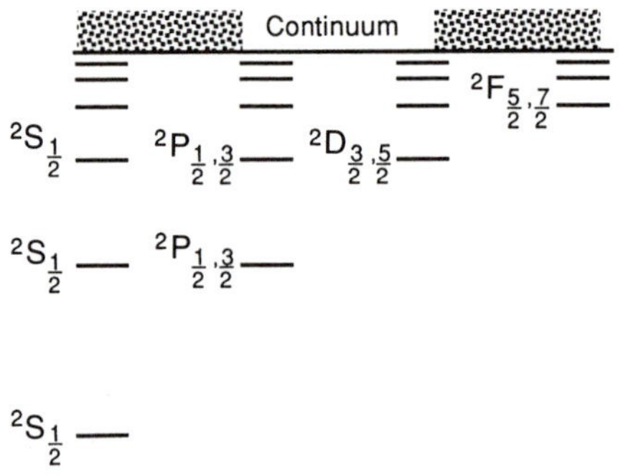

Continuum

$^2\mathrm{S}_{\frac{1}{2}}$ $^2\mathrm{P}_{\frac{1}{2},\frac{3}{2}}$ $^2\mathrm{D}_{\frac{3}{2},\frac{5}{2}}$ $^2\mathrm{F}_{\frac{5}{2},\frac{7}{2}}$

$^2\mathrm{S}_{\frac{1}{2}}$ $^2\mathrm{P}_{\frac{1}{2},\frac{3}{2}}$

$^2\mathrm{S}_{\frac{1}{2}}$

Fig. 1.23 Term symbols for the hydrogen atom energy levels.

The energy level diagram for the hydrogen atom is shown in Fig. 1.23, with the correct term symbols; note that the splittings between pairs of terms from the same electron configuration are too small to be visible on this diagram.

Lithium

The electron configuration of the ground state of the Li atom is $1s^2\ 2s^1$. The 1s shell is completely full, so as we saw above, makes no contribution to S, L, or J; as long as we do not excite an electron out of the $1s^2$ core, the problem is effectively a one-electron case, and the terms resemble those seen in hydrogen.

For the ground state, $L = 0$, $S = \frac{1}{2}$, and so $J = \frac{1}{2}$; the term symbol is $^2S_{\frac{1}{2}}$. For the excited states $1s^2$ 3s, $1s^2$ 4s, etc, we again obtain a series of $^2S_{\frac{1}{2}}$ states, each having a higher energy than the previous one. This is illustrated in Fig. 1.24; the main difference from the hydrogen diagram is that, in Li, there is no simple relationship between the energies of the various terms.

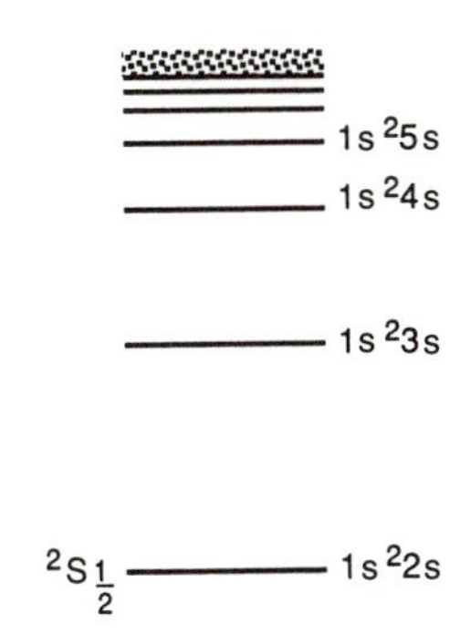

Fig. 1.24 $^2S_{\frac{1}{2}}$ levels of the lithium atom.

There is also a series of excited states $1s^2$ 2p, $1s^2$ 3p, etc; just as in the hydrogen case, these each give rise to two terms, $^2P_{\frac{3}{2}}$ and $^2P_{\frac{1}{2}}$. In the absence of magnetic effects, these would be degenerate; there is in fact a small splitting between them, just as there was in hydrogen. The $1s^2np$ states are not degenerate with the corresponding $1s^2ns$ states, as they are in hydrogen; as we saw earlier, they lie at rather high energies.

Terms similarly arise from configurations such as $1s^2$ nd, $1s^2$ nf, and so on; for a given value of n, their energies increase steadily as l increases.

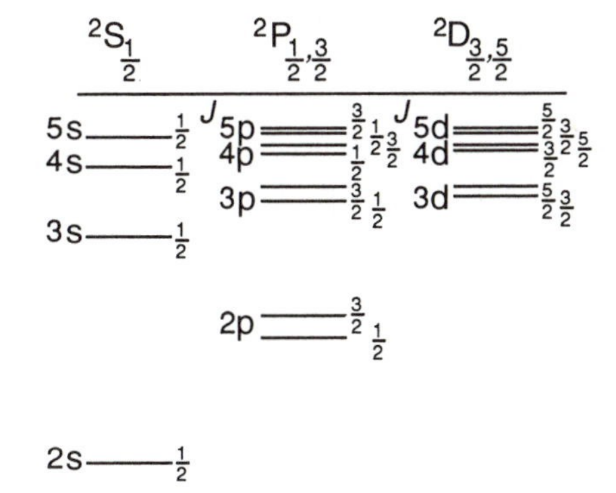

Fig. 1.25 Energy levels of the lithium atom (not to scale).

The complete energy level diagram for Li is shown in Fig. 1.25. The splittings between terms from the same configuration are exaggerated for clarity. Configurations in which an electron is excited from the $1s^2$ core are of much higher energy, and do not appear in this diagram.

Helium

The He atom has a ground state configuration $1s^2$, which is a closed-shell configuration; as we have seen, its term symbol is $^1S_0$.

The simplest excited state has the configuration 1s 2s, and this gives rise to more than one term. When both electrons were in a 1s orbital, the Pauli exclusion principle required that they had opposed spins, so that they did not have all four quantum numbers the same. Now the electrons already differ in their principal quantum number, and so the spins may either be parallel or anti-parallel. The value of the orbital quantum number L is simply found, as both electrons are in s orbitals:

$$l_1 = 0 \text{ and } l_2 = 0, \text{ therefore } L = 0.$$

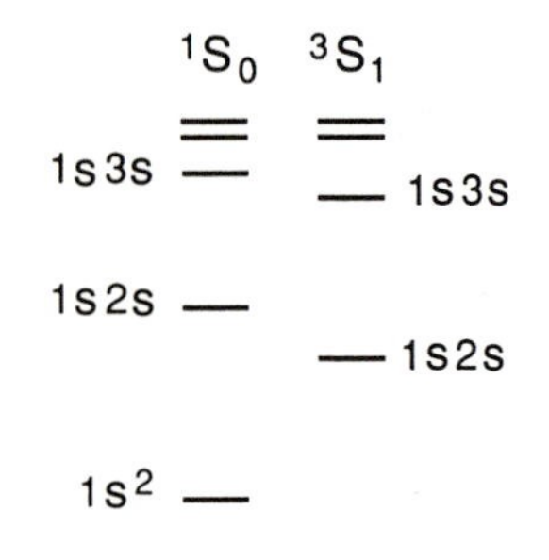

Fig. 1.26 S levels of helium.

For the spin quantum number S,

$$s_1 = \tfrac{1}{2} \text{ and } s_2 = \tfrac{1}{2},$$

therefore $S = 0$ or 1, and so $J = 0$ or 1. There are therefore two terms, $^1S_0$ and $^3S_1$ corresponding to the two electron spins being anti-parallel or parallel respectively. These terms are not degenerate, and the triplet state, with the electron spins parallel, is of lower energy; this is in agreement with Hund's rule, which we saw earlier. The same argument holds for other excited configurations 1s 3s, 1s 4s, and so on, so we can draw part of the energy level diagram as two separate ladders, as in Fig. 1.26.

There are further excited states with configurations of the type 1s np, and for these:

$$l_1 = 0 \quad l_2 = 1 \ \therefore \ L = 1$$
$$\text{and } s_1 = \tfrac{1}{2} \quad s_2 = \tfrac{1}{2} \ \therefore \ S = 0 \text{ or } 1.$$

If we consider the $S = 0$ case, then as $L = 1$, $J = 1$, and the term symbol is $^1P_1$. In the $S = 1$ case, then there are three possible values of J, 0, 1, and 2, as shown in Fig. 1.27. The term symbols are therefore $^3P_0$, $^3P_1$, and $^3P_2$. As we have seen before, the splittings between terms which differ just in their values of J are very small—as they are magnetic in origin—but the splitting between the singlet and triplet states—which is electrostatic in origin—is significantly greater. The part of the energy level diagram corresponding to the 1s np configurations is shown in Fig. 1.28.

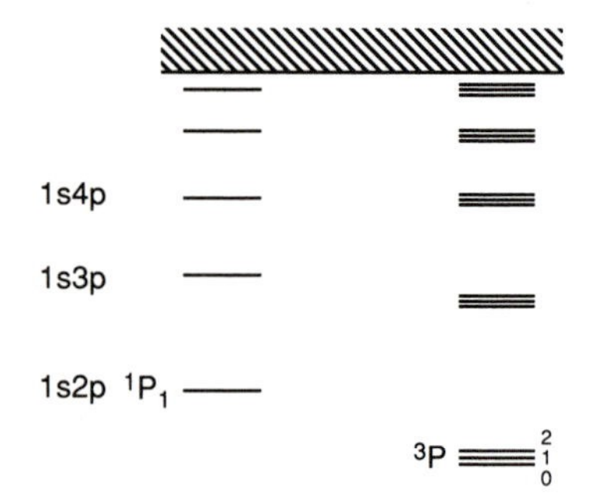

Fig. 1.28 The P states of He.

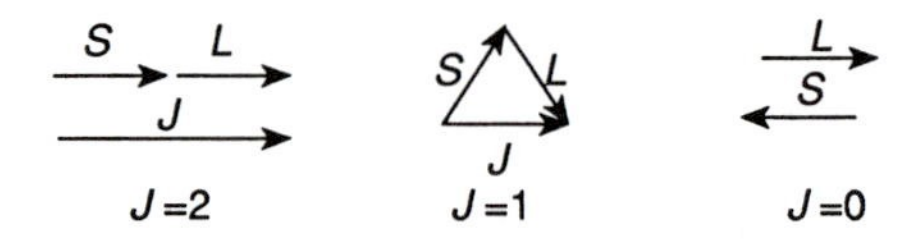

Fig. 1.27 Coupling of L and S for ^3P states of He.

In the same way, we can show that the configuration 1s 3d gives rise to term symbols $^1D_2$ (when the spins are anti-parallel), and $^3D_1$, $^3D_2$, and $^3D_3$ (when the spins are parallel).

1.23 He as a model for photochemistry

The energy levels of the He atom provide a simple model for the changes which are common in photochemistry.

Most stable molecules, particularly in organic chemistry, have all their electrons in closed shells; the electrons are paired, with the spin of one being cancelled by the spin of its partner. The ground states of these molecules therefore have no resultant spin angular momentum, and normally they have no orbital angular momentum either. Their ground states are therefore singlets, with $S = 0$.

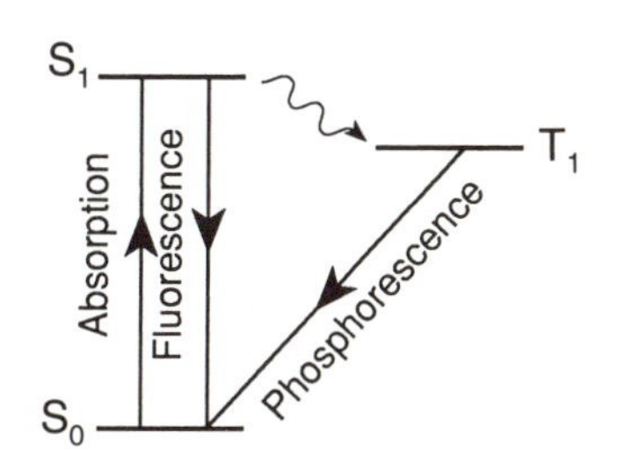

Fig. 1.29 Photochemical processes in an organic molecule.

Excitation by light will split one electron pair, leaving the other pairs substantially unchanged. The split electron pair is therefore similar to the electron pair in the He atom, although the situation is usually simpler, as there is no orbital angular momentum to take into account. We can therefore visualize a three-state system, with a ground state which is a singlet, and two excited states, a singlet and a triplet, corresponding to the unpaired electrons having parallel or anti-parallel spins. This is shown in Fig. 1.29, where the ground state is labelled S_0, the excited singlet state S_1, and the excited triplet state T_1. In accordance with Hund's rule, T_1 has a slightly lower energy than S_1.

Photon absorption causes the molecule to be excited from S_0 to S_1; a direct transition to T_1 is not allowed by spectroscopic selection rules, as it involves a change in the quantum number S. The molecule may then be able to re-radiate its energy, falling back to the ground state S_0. This typically takes times of the order of 10^{-8} seconds, and is called *fluorescence*. It is also possible for the state T_1 to be populated directly from S_1 by *inter system crossing*. From T_1 it is now possible for the molecule to return to the ground state S_0; this transition is of very low probability, as it is in principle forbidden by a selection rule; it involves a change in the quantum number S. In practice it does occur slowly, taking times of the order of 1 second; there is often no other possible fate for the molecule, as there is no other triplet of lower energy. This delayed emission of light is known as *phosphorescence*.

The chemical importance of photochemistry is often dependent on the long lifetime of molecules in triplet states. Such molecules can take part in striking chemical reactions; they have a lifetime which is long compared with those required for molecular collisions or rearrangements.

1.24 Spin–orbit coupling

We have seen that a splitting is observable in the ^{2}P states of the hydrogen atom, and of the alkali metals such as lithium. The energy of the ^{2}P state depends on the relative orientation of the spin and orbital angular momenta, described by L and S; the effect is called spin–orbit coupling, and is caused by the spinning electron acting as a tiny magnet which interacts with the field caused by its motion around the nucleus.

The magnitude of the spin–orbit splitting is roughly dependent on Z^4, where Z is the nuclear charge; the splittings therefore increase very rapidly down the Periodic Table. This is illustrated in Fig. 1.30, which gives the splittings between the lowest $^2P_{\frac{3}{2}}$ and the $^2P_{\frac{1}{2}}$ levels in the elements of Group 1. The splitting is much larger in Cs than Li, but even in Cs the splitting is smaller than a typical energy gap between two different electron configurations. The splitting of the familiar D lines in the emission spectrum of Na is caused by the separation of 17.2 cm^{-1} between the lowest ^{2}P levels.

Spin–orbit coupling can allow an atom to undergo a transition which would otherwise be forbidden by selection rules, such as a transition from a singlet to a triplet state.

It is often said that the coupling of spin and orbital angular momenta mixes the multiplicities of terms, and so, for example, singlet terms acquire some triplet nature, and vice versa. L and S are then said not to be 'good' quantum

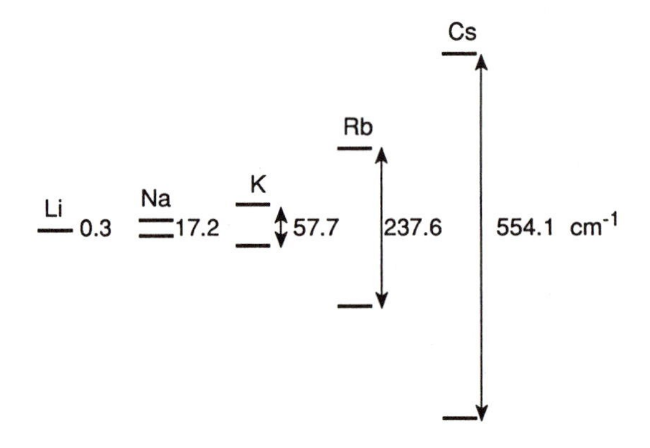

Fig. 1.30 Splittings of the $^2P_{\frac{3}{2}}$ and $^2P_{\frac{1}{2}}$ levels of the alkali metals.

numbers. Of course, if the spin–orbit coupling is small, then the description of terms by their values of L and S may still be a very close approximation, even if they are not completely accurate.

1.25 jj coupling

In this chapter we have drawn energy level diagrams based on the Russell–Saunders scheme, in which the electrostatic interactions between electrons dominate, and the magnetic interactions can be thought of as small perturbations. This description holds well for light atoms, but very heavy atoms are best treated by a different scheme, jj coupling, in which magnetic effects are much more important.

In jj coupling, the rules for coupling angular momenta are:

1. For each electron i, the orbital and spin angular momenta l_i and s_i combine to give a resultant total angular momentum for the electron j_i.

2. The individual values of j_i then couple to give a resultant J, the total angular momentum for the atom.

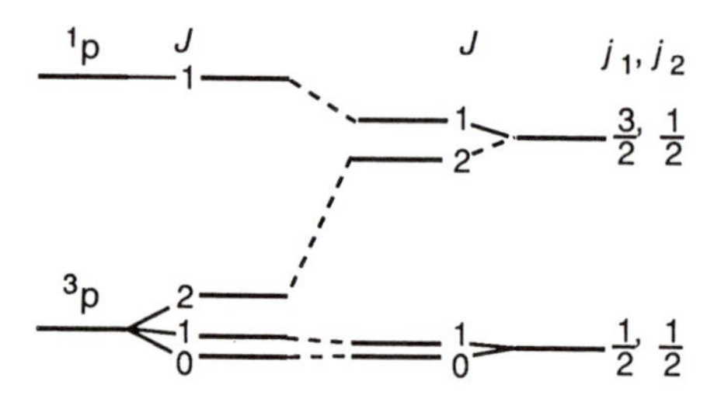

Fig. 1.31 The transition from Russell–Saunders to jj coupling.

This can be illustrated by considering a state with one s and one p electron. In this case,

$$l_1 = 0 \quad s_1 = \tfrac{1}{2} \ \therefore \ j_1 = \tfrac{1}{2}$$
$$l_2 = 1 \quad s_2 = \tfrac{1}{2} \ \therefore \ j_2 = \tfrac{3}{2} \text{ or } \tfrac{1}{2}.$$

If $j_2 = \frac{1}{2}$, then $J = 0$ or 1; if $j_2 = \frac{3}{2}$, then $J = 1$ or 2. There are therefore four different terms, each described by their values of j_1, j_2, and J, as shown in Fig. 1.31. The values of L and S are not well defined.

This is exactly the same number of terms that we derived for the 1s 2p excited state in He, under the Russell–Saunders scheme. As Fig. 1.31 shows, the relative energies of the terms were then rather different, but the number of states, and their J values, are identical. This reminds us that Russell–Saunders and jj coupling are merely two extreme models; reality is usually somewhere in between these two extremes, and we must use whichever scheme gives the better description of an atom. As we have seen, in general the Russell–Saunders scheme is more appropriate for light atoms, and jj coupling for heavier atoms.

2 Diatomic molecules

2.1 The orbital approximation

The orbital approximation, so useful in understanding the electronic structure of atoms, carries over directly to molecules. Thus for a molecule the wave function is expressed approximately as a product of one-electron wave functions, each with spin α or β; electronic structures are then written in the following way:

$$\begin{aligned} \mathrm{H_2} &\quad : \quad 1\sigma_g^2 \\ \mathrm{CO} &\quad : \quad 1\sigma^2\, 2\sigma^2\, 3\sigma^2\, 4\sigma^2\, 1\pi^4\, 5\sigma^2. \end{aligned}$$

Many aspects of the atomic notation carry over to the molecular situation. The running numbers 1, 2, 3, etc. increase as the energy of the orbital increases, and they are like the atomic quantum number n. The symbols σ, π, etc. describe the symmetry of the one-electron functions for the molecule in a manner parallel to that coded as s, p, d, f, etc. for atoms. The superscripts give the number of electrons in each orbital, and, in the examples above, all the electrons are paired with partners of opposite spin so that they have closed shell or inert-gas-like stability. The overwhelming number of molecules dealt with by chemists do have closed-shell ground states, although a few molecules, such as NO and O_2, do have unpaired electrons.

The Pauli principle applies to electrons in molecules as well as in atoms. In its simplest form, it prevents any pair of electrons in a molecule having identical quantum numbers—which is why all 14 electrons in CO do not occupy the 1σ orbital. In its more fundamental form, which we saw in Section 1.11, it requires that the wave function must be anti-symmetric to electron exchange. This means that in the H_2 molecule the configuration $1\sigma_g^2$ is a shorthand form of the full wave function:

$$\Psi = 1\sigma_g(\alpha)\,(1)\; 1\sigma_g(\beta)\,(2) - 1\sigma_g(\alpha)\,(2)\; 1\sigma_g(\beta)\,(1).$$

This wave function is in fact the determinant of a matrix:

$$\Psi = \begin{vmatrix} 1\sigma_g(\alpha)\,(1) & 1\sigma_g(\alpha)\,(2) \\ 1\sigma_g(\beta)\,(1) & 1\sigma_g(\beta)\,(2) \end{vmatrix}$$

and all wave functions can be anti-symmetrized by forming similar determinants.

2.2 The Born–Oppenheimer approximation

An added complication possessed by molecules, but not by atoms, is that the energy of a molecule depends on the relative positions of its nuclei. Even without exciting electrons to higher orbitals, the energy of the H_2 molecule will vary as the H–H bond is stretched or compressed.

The Born–Oppenheimer approximation allows us to handle this extra complication. The approximation treats nuclear and electronic motion as entirely independent; this is reasonable, as the velocity of nuclear motion is very much less than that of the much lighter electrons.

In practice this means that we consider a molecule with a given nuclear framework, and calculate its electronic energy only. The nuclear–nuclear repulsion is then added on as a separate term, and is calculated as a simple sum of Coulomb repulsion energies. We may then repeat the process for a different set of coordinates of the nuclei, and find a new energy. This procedure is the basis for drawing potential curves or surfaces. In Fig. 2.1 the potential curve shows how the energy of H_2 varies with the internuclear distance.

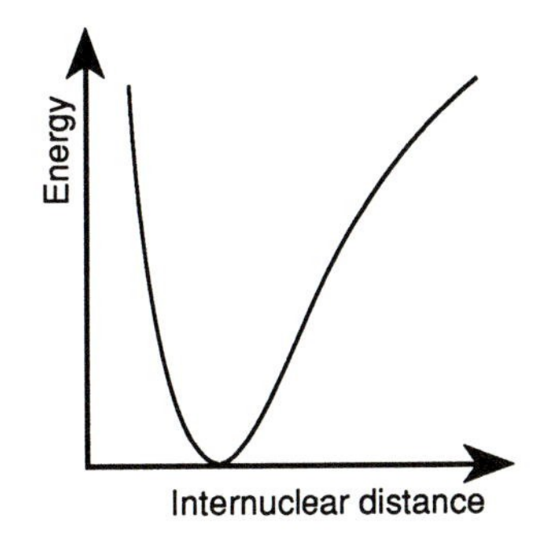

Fig. 2.1 The potential-energy curve for a diatomic molecule.

2.3 The LCAO approximation

The total wave function for a molecule has now been separated into two parts, one part describing the nuclei and the other the electrons. The electronic part of the wave function is written in the orbital approximation as:

$$\Psi = \Phi_1 \, \Phi_2 \ldots \Phi_n,$$

where each Φ represents the wave function of a single electron, and the wave function is anti-symmetrized to take account of the Pauli principle. The task for the theoretical chemist is now to discover what forms the orbitals Φ take.

One of the commonest approaches to this problem is to express each orbital as a sum of atomic orbitals (1s, 2p, etc.) based on the various atoms in the molecule. This is called the Linear Combination of Atomic Orbitals (LCAO); each orbital may be written

$$\Phi = \sum_m c_m \, \chi_m,$$

where χ_m is an atomic orbital and c_m is a mixing coefficient. Provided that enough atomic orbitals are considered, this can be a very accurate method of representing molecular orbitals. The values of mixing coefficients c_m are adjusted so that the energy of the molecule, as calculated by the Schrödinger equation is as low as possible; a theorem known as the variation principle shows that the lower the calculated energy of the approximate wave function, the more accurately it represents the true wave function.

The calculation of molecular orbitals as linear combinations of many atomic orbitals presents massive computational problems, and is feasible only for small molecules. However, the essential qualitative features of chemical bonding can be understood simply, if the atomic orbitals from which the molecular orbitals are constructed are restricted just to those which are occupied in the separate atoms.

2.4 Bonding and anti-bonding molecular orbitals

Consider the hydrogen molecule, H_2. It contains two electrons; if the H–H bond were broken, we would have two hydrogen atoms, each with one electron in a 1s orbital. These 1s orbitals are identical to each other and have the form $\chi = Ne^{-r}$. We shall label them $1s_A$ and $1s_B$, with A and B representing the two nuclei, and use just these two orbitals to construct the molecular orbitals of the hydrogen molecule.

There is one further feature of the H_2 molecule that we can use: as the molecule is symmetrical, we would expect the electron density to be the same round each nucleus. As the electron density is proportional to the wave function squared, this means that the wave function must either be identical round each nucleus or be the same except for a change of sign.

We can now express our molecular orbitals Φ in terms of the atomic orbitals $1s_A$ and $1s_B$:

$$\Phi = c_A 1s_A + c_B 1s_B.$$

One way in which we can ensure that the electron density is the same round each nucleus is by taking $c_A = c_B$; if we are not concerned with normalizing our function, then we can put both coefficients equal to one, so that

$$\Phi = 1s_A + 1s_B,$$

which is illustrated pictorially in Fig. 2.2. This molecular orbital is called the $1\sigma_g$ orbital; it is a three-dimensional function which is positive everywhere.

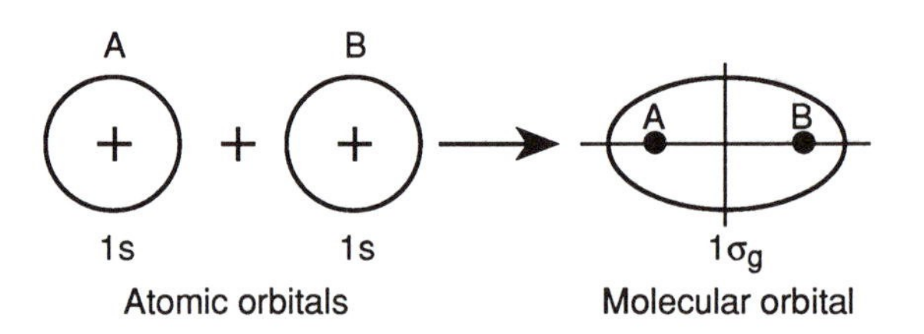

Fig. 2.2 Formation of the $1\sigma_g$ molecular orbital.

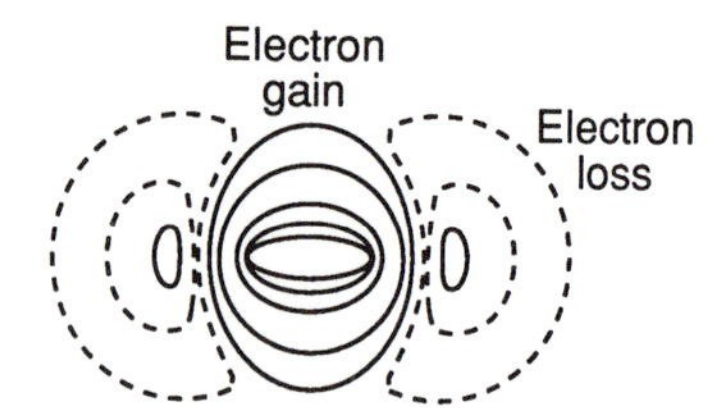

Fig. 2.3 Electron-density difference between H_2 and two individual H atoms with electrons in a σ_g molecular orbital.

The notation σ indicates that it is cylindrically symmetrical about the A–B axis. The subscript g stands for the German word *gerade*, even, meaning that the function is also symmetric with respect to the centre of symmetry. Figure 2.3 shows how the electron density, which is proportional to Φ^2, compares with the electron density in two separate hydrogen atoms. The formation of the molecular orbital produces a build-up of negative charge between the positive nuclei; this is energetically favourable and so the $1\sigma_g$ orbital is called a *bonding* orbital.

There is one other way of combining the $1s_A$ and $1s_B$ orbitals so as to produce an electron density which is the same round each nucleus. If $c_A = -c_B$, then the sign of the wave function is positive round one nucleus and negative round the other, but the electron densities, which depend on Φ^2, are the same. Thus

$$\Phi = 1s_A + 1s_B,$$

This molecular orbital is called $1\sigma_u$ and is shown in Fig. 2.4, where again σ indicates cylindrical symmetry, while the subscript u (from the German word for odd, *ungerade* refers to the function being antisymmetric—changing sign on reflection at the centre of symmetry.

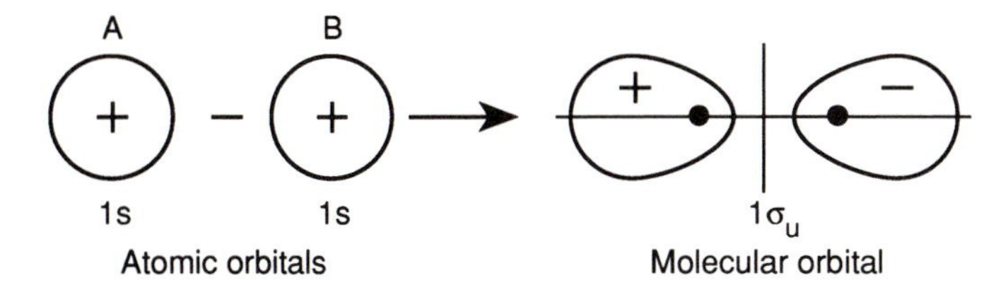

Fig. 2.4 Formation of the $1\sigma_u$ molecular orbital.

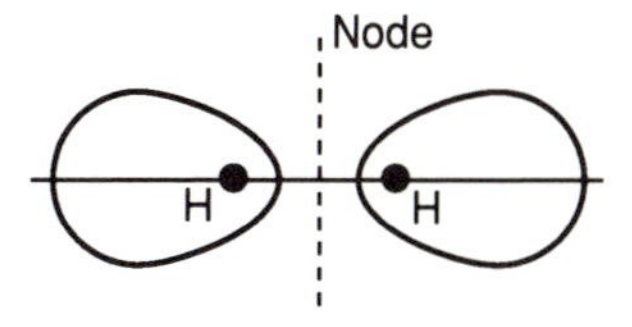

Fig. 2.5 Electron distribution of the $1\sigma_u$ orbital of H_2.

The electron density of the σ_u orbital is shown in Fig. 2.5; the charge build-up is now away from the positively charged nuclei. This is energetically unfavourable, and the σ_u orbital is called an *anti-bonding* orbital. Note that it is still axially symmetrical and is therefore still a σ orbital.

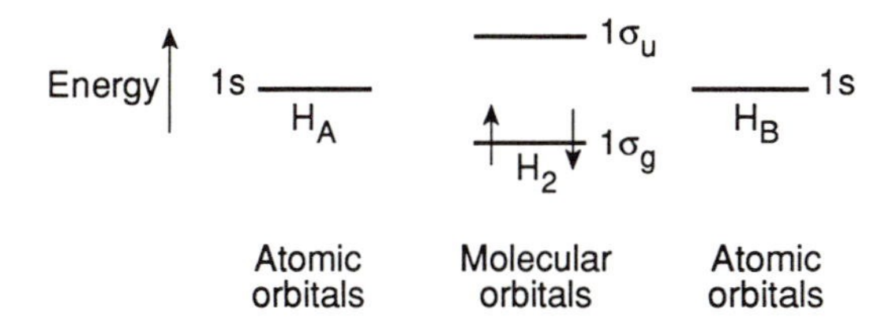

Fig. 2.6 Molecular orbital diagram for H_2.

The formation of two molecular orbitals when two hydrogen atoms come together is illustrated in Fig. 2.6; the separate atomic orbitals are shown on the left and right, and the molecular orbitals in the centre. In the H_2 molecule both electrons can occupy the bonding $1\sigma_g$ orbital, and the molecule is therefore more stable than two separate hydrogen atoms.

The amount by which the $1\sigma_g$ orbital is stabilized depends on how much the two hydrogen atomic orbitals overlap. If the distance between the nuclei is large, then there is little overlap and the molecule is not very stable. The stability of the electrons increases as the nuclei come closer together. However, the total energy of the molecule is the sum of the electronic energy and the nuclear–nuclear repulsion energy; this repulsion energy dominates at very short

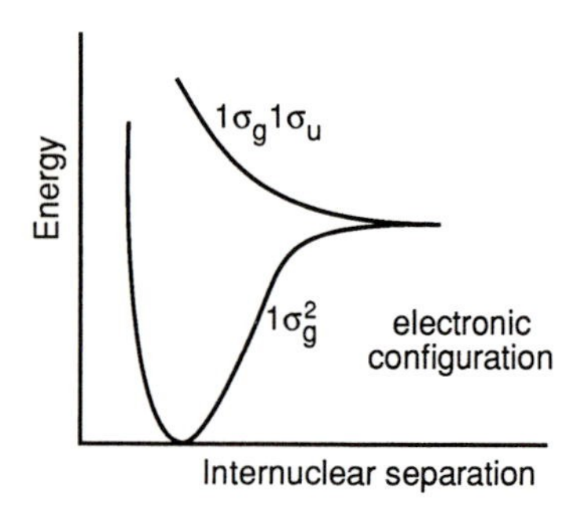

Fig. 2.7 Potential-energy curves for the hydrogen molecule in its ground state.

distances, and so the variation of the total energy with bond length for an H_2 molecule is shown in Fig. 2.7.

Promoting an electron to the excited $1\sigma_u$ molecular orbital will result in one bonding and one antibonding electron. There is now no net bonding, and the resulting molecular electronic state is unstable.

Table 2.1 Bonding in H_2 and H_2^+

	H_2	H_2^+
Bond length (nm)	0.074	0.11
Bond energy (kJ mol^{-1})	436	260
Bonding electrons	2	1

The molecular orbital diagram in Fig. 2.6 can also be used to describe the structure of the ion H_2^+; although this ion is not very familiar, it has been observed in the gas phase, and is stable to dissociation. It contains just one electron, which occupies the bonding $1\sigma_g$ orbital; Table 2.1 shows how the number of bonding electrons affects the physical properties of H_2 and H_2^+.

2.5 Molecular orbital diagrams

The idea that overlapping atomic orbitals can form bonding and anti-bonding orbitals can be extended easily to more complex diatomic molecules.

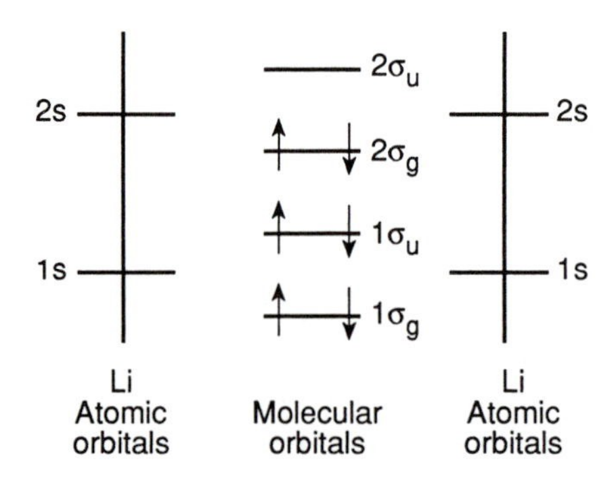

Fig. 2.8 The molecular orbital diagram for Li_2.

The molecule Li_2 is a stable molecule found in the vapour above heated Li metal. In the case of the Li_2 molecule, the fully occupied 1s orbitals hardly overlap at all, and make no contribution to the bonding. The 2s orbitals do overlap, however, and form two molecular orbitals, $2\sigma_g$ and $2\sigma_u$, which are respectively bonding and anti-bonding. The two remaining electrons therefore occupy the bonding orbital, as shown in Fig. 2.8, and form a single bond.

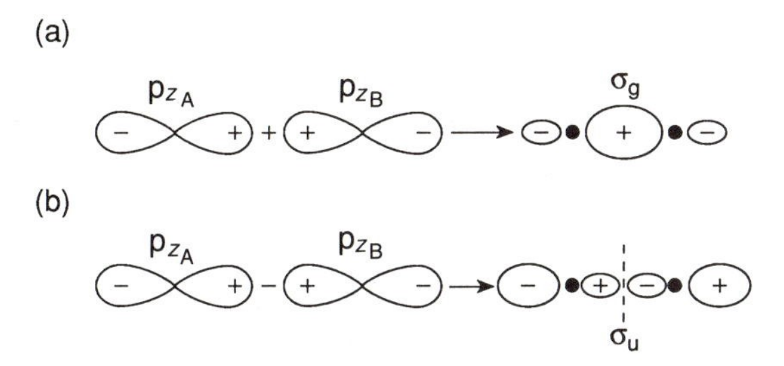

Fig. 2.9 Molecular orbitals from atomic $2p_z$ functions.

For the molecule N_2 we need to consider the overlap of 2p as well as 2s orbitals; these similarly form bonding and anti-bonding molecular orbitals. It is conventional to describe the internuclear axis as the z axis, so that the $2p_z$ atomic orbitals give two molecular orbitals which are cylindrically symmetrical about the z-axis (Fig. 2.9). The $2p_x$ orbitals can also overlap, but these form molecular orbitals which are not symmetrical about the bond axis, and are called π orbitals (Fig. 2.10). The $2p_y$ atomic orbitals form a similar pair of molecular orbitals at right angles to those from the $2p_x$ orbitals; the two bonding orbitals have identical energies, and are said to be degenerate.

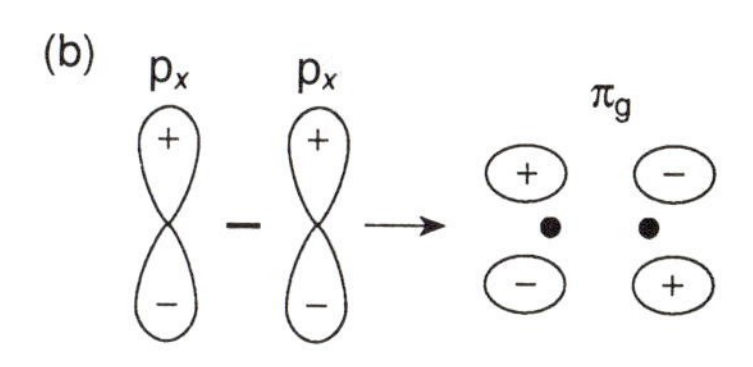

Fig. 2.10 Linear combinations of atomic $2p_x$ orbitals.

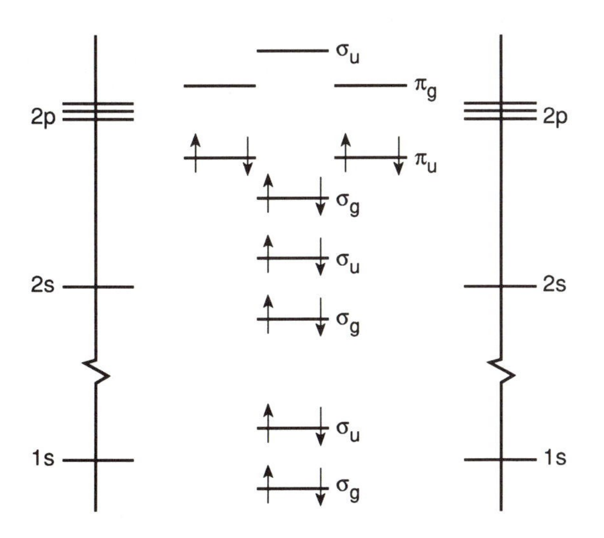

Fig. 2.11 The molecular orbital diagram for N_2.

Figure 2.11 now shows the molecular orbital diagram for N_2; the 1s orbitals have been omitted for clarity. There are ten electrons outside the 1s cores; four of these occupy the bonding and anti-bonding orbitals from the 2s orbitals—producing no net bonding—and the remaining six occupy the three bonding orbitals from the 2p orbitals. There is therefore an excess of six bonding electrons over anti-bonding electrons, and this corresponds to a triple bond.

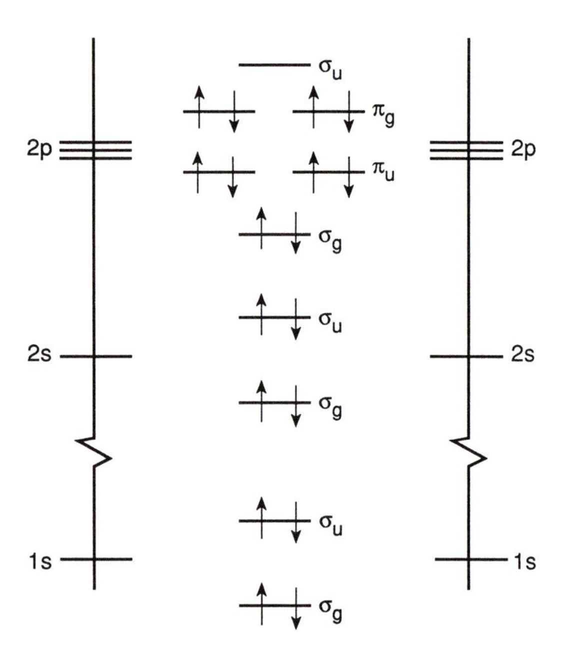

Fig. 2.12 The molecular orbital diagram for F_2.

The same molecular orbital diagram will also describe the molecules F_2 and O_2; we simply need to feed in different numbers of electrons. Figure 2.12 shows the diagram for F_2; the four extra electrons now occupy the two anti-bonding orbitals. This means that in F_2 there are eight bonding electrons and six anti-bonding electrons; the excess of two bonding electrons corresponds to a single covalent bond.

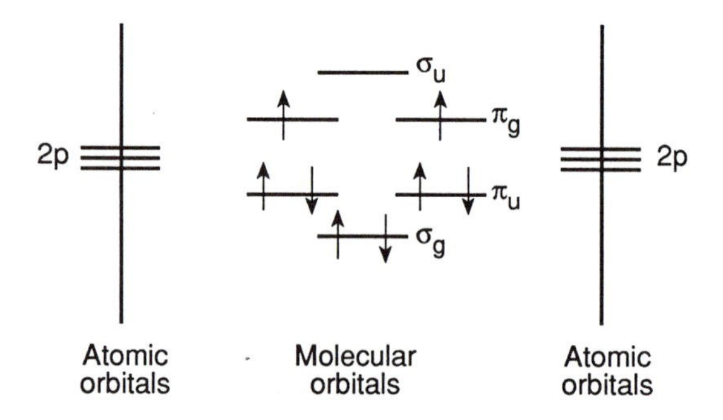

Fig. 2.13 The higher occupied molecular orbitals in O_2.

The case of O_2 is a little more complex, as there are two degenerate π_g orbitals available for the last two electrons. The most stable arrangement is shown in Fig. 2.13; the electrons occupy different orbitals, to minimize electron repulsion, and have parallel spins. This is an example of Hund's rule, which we first encountered in atomic structure. As a consequence of this, O_2 molecules

Table 2.2 Bonding in O_2 and O_2^+

	O_2	O_2^+
Bond length (nm)	0.121	0.112
Bonding electrons	$6 - 2 = 4$	$6 - 1 = 5$

have a resultant spin angular momentum, which gives rise to the paramagnetism of liquid oxygen.

In O_2 the electron of highest energy is in an anti-bonding orbital; it follows that the removal of this electron will make the bonding between the nuclei stronger. Table 2.2 compares the physical properties of O_2 and O_2^+; this can be contrasted with the data for H_2 and H_2^+ in Table 2.1, where the electron removed is a bonding electron.

So far the molecules that we have considered have had identical nuclei, that is they are homonuclear. For a heteronuclear molecule such as CO, molecular orbitals can still be constructed from atomic orbitals, but now the coefficients of these atomic orbitals will no longer be determined by symmetry. Molecular orbitals can vary from sharing electrons almost equally between two atoms, to being largely localized on one atom. Orbitals are still described as σ or π, but the distinction between u and g now disappears, as the molecule has no centre of symmetry. A simple molecular orbital diagram for CO is shown in Fig. 2.14.

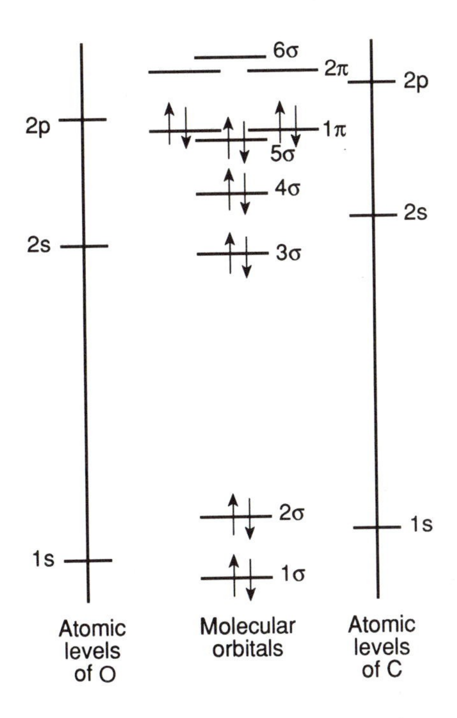

Fig. 2.14 The molecular orbital diagram for CO.

2.6 Designation of diatomic electronic energy levels

In the case of atoms we saw that the angular momenta of individual electrons coupled together to produce resultants which serve to label an electronic state; a similar situation applies in molecules.

A molecule is described by its total spin and orbital angular momenta; these can be obtained by adding up the spin and orbital angular momenta of the individual electrons. Just as in atoms, this procedure is much simplified by the fact that all completely filled orbitals have no resultant spin or orbital angular momentum; in most cases we have only a few electrons in partially filled orbitals to consider.

The spin angular momentum of a molecule is described by the quantum number Σ, and just as in atoms, is often coded as the multiplicity, $2\Sigma + 1$. Thus for a closed shell molecule, $\Sigma = 0$, the multiplicity is 1, and the state is described as a singlet. For a molecule with just one unpaired electron, $\Sigma = \frac{1}{2}$, the multiplicity is 2, and the state is a doublet.

The orbital angular momentum about the internuclear axis is described by the quantum number Λ. σ electrons have no orbital angular momentum, and so contribute nothing to Λ. π electrons contribute one unit of angular momentum; when we have two degenerate π orbitals, one contributes $+1$, and the other -1. (More strictly we should refer to the linear combinations π^+ and π^-, which are $(\pi_x + \mathrm{i}\pi_y)$ and $(\pi_x - \mathrm{i}\pi_y)$ respectively.)

The term symbols for molecules are written in a similar way to those for atoms; the multiplicity is written as a superscript, and the value of Λ is coded Σ for 0, Π for 1 and Δ for 2. Thus a closed-shell molecule would be described as $^1\Sigma$, and other states might be written $^2\Pi$ or $^3\Sigma$. We shall see examples of how term symbols are related to electron configurations in the next section.

Two further additions may be made to the term symbol. For homonuclear diatomics the subscript g or u is added, to indicate whether the total wave function is symmetric or antisymmetric with respect to inversion at the origin. States are u if the number of u orbitals is odd, and g if the number of u orbitals is even.

The other addition refers to the symmetry of the wave function with respect to the plane through the nuclei, described by a superscript plus or minus. This symmetry leaves σ orbitals unchanged, but converts π^+ to π^- and vice versa. Evaluating the effect of this operation on the whole molecular wave function requires care, as wave functions are really determinants; exchanging two columns in a determinant changes the sign of the determinant. The ground states of N_2 and H_2 can hence be shown to be $^1\Sigma_g^+$, while that of O_2 is $^3\Sigma_g^-$.

2.7 Molecular electronic energy levels

We can now apply the ideas of the last two sections to obtain descriptions of the ground and excited states of some simple molecules.

Hydrogen

The ground state configuration of the H_2 molecule is $1\sigma_g^2$. The total orbital angular momentum is zero, as both electrons are in σ orbitals and the total spin is zero, and as the electrons are in the same orbital and so must have opposite spins. The wave function is unchanged on inversion at the centre of the molecule, or reflection in the internuclear axis; the ground state is therefore $^1\Sigma_g^+$.

If we promote one of the electrons to the anti-bonding $1\sigma_u$ orbital, the configuration becomes $1\sigma_g\ 1\sigma_u$. The orbital angular momentum is still zero, and the wave function is still unaltered on reflection in the internuclear axis. However there is now one electron in a u orbital, so the total wave function is u. As the electrons occupy different orbitals, there is no restriction on the spins of the electrons; they may be parallel, so $\Sigma = 1$, or anti-parallel, with $\Sigma = 0$. There are therefore two excited states from this electron configuration, $^1\Sigma_u^+$ and $^3\Sigma_u^+$. Neither of these states is bound, that is to say there are no minima in their potential-energy curves.

The hydrogen chloride ion

The ion HCl^+ might seem an obscure example to illustrate electronic energy levels; in fact the ion is formed from HCl in photoelectron spectroscopy, and an understanding of its ground and excited states is essential to the interpretation of the photoelectron spectrum of HCl.

Figure 2.15 shows the molecular orbital diagram for the valence electrons in HCl. The hydrogen 1s orbital and the chlorine $2p_z$ orbital overlap forming a bonding and an anti-bonding orbital; the two remaining 2p orbitals are non-bonding. The most stable configuration for HCl^+ is therefore $\sigma^2\ \pi^3$ and there is a more excited configuration $\sigma^1\ \pi^4$. Both configurations have one unpaired electron, and so $\Sigma = \frac{1}{2}$; they are both doublets. The more stable state has a total of one unit of orbital angular momentum from its π electrons, and is therefore a $^2\Pi$ state; the excited state has no orbital angular momentum, as its π orbitals are full, and is therefore a $^2\Sigma$ state.

There is one further complication in the $^2\Pi$ state of HCl^+. The ion has spin angular momentum of $\frac{1}{2}$, and orbital angular momentum of 1; these can couple together to give a total angular momentum of $\frac{3}{2}$ or $\frac{1}{2}$, depending on their relative orientations. These two states have energies which are slightly different from each other, due to spin–orbit coupling; this effect was also encountered in the structure of atoms.

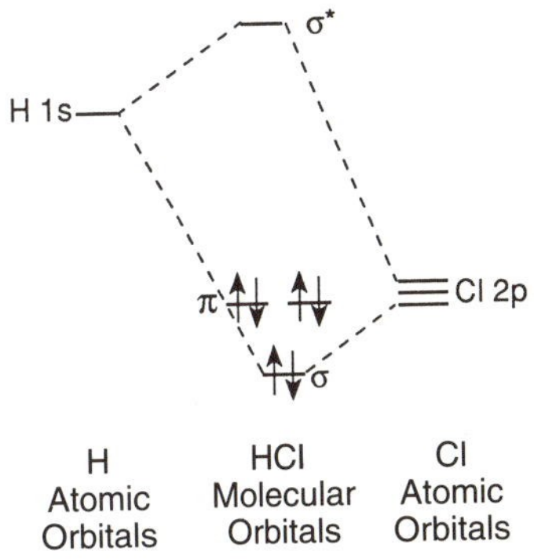

Fig. 2.15 The higher occupied orbitals in HCl.

Oxygen

We have seen already that the electron configuration of the O_2 molecule is determined by Hund's rule; the two electrons of highest energy occupy two different degenerate π orbitals, with their spins parallel. Although this is the most stable arrangement, it is not the only one possible, and other excited states are obtained from redistributing the electrons in the π_g orbital, as shown in Fig. 2.16.

The ground state has the configuration $\pi_g^+\ \pi_g^-$, if we ignore the more stable filled orbitals, with both electrons having the same spin. The total electron spin

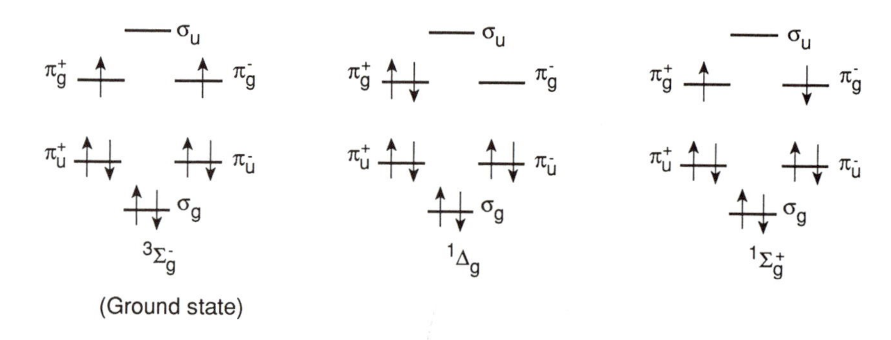

Fig. 2.16 Occupation of orbitals in states of O_2.

is therefore one, but the orbital angular momentum is zero, as the contributions from the two π orbitals cancel each other. There are no u electrons, so the total wave function is g; it is also anti symmetric to reflection in the internuclear axis, so the state is written as $^3\Sigma_g^-$.

An excited state is obtained by keeping the electrons in the same orbitals, but giving them opposed spins. Once again the orbital angular momentum is zero, but now the total spin is also zero; the state is now written $^1\Sigma_g^+$.

A second excited state is obtained if we put both electrons in the same orbital. Now they must have opposite spins, so $\Sigma = 0$, but their orbital angular momenta now add up to 2; the state is therefore described as $^1\Delta$. Further excited states are obtained by promoting electrons to higher anti-bonding orbitals; these are of even higher energies.

2.8 Vibrational energy levels

We saw in Section 2.2 that the energy of a molecule depends on the internuclear distance, and that we can describe this variation by a potential-energy curve. This means that a molecule can vibrate, and not surprisingly we find that its vibrational energy is quantized.

The simplest model to describe the vibration of a molecule is the harmonic oscillator, in which the restoring force is proportional to the displacement of the nuclei from the equilibrium position. This is equivalent to thinking of the molecule as two weights connected by a spring, and the potential curve has the shape of a parabola. The Schrödinger equation can be solved exactly for this case, and the potential energy curve and resulting vibrational energy levels are shown in Fig. 2.17.

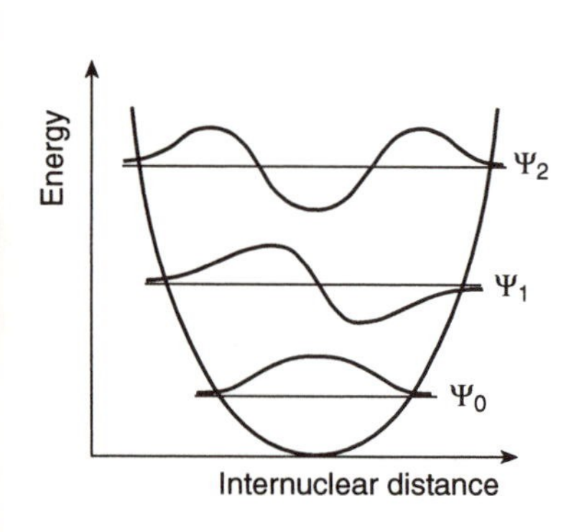

Fig. 2.17 Energy levels and wave functions of a harmonic oscillator.

The energy levels are described by the quantum number v, which takes the values 0, 1, 2, etc., and are equally spaced. The energy of the levels is given by the equation:

$$E = (v + \tfrac{1}{2})\, h\nu$$

where ν is the vibration frequency. Note that even the lowest energy level, when $v = 0$, involves some vibrational energy, the so-called zero-point energy; this is a consequence of the Uncertainty Principle.

A real diatomic curve, although being very similar to the harmonic oscillator at the bottom of the well, diverges at higher energies. The left-hand side increases in energy indefinitely, reflecting the difficulty of forcing the two nuclei closer and closer. However the right-hand side, corresponding to large internuclear separation, flattens out to the point where the bond is broken. The result of this anharmonicity is to cause the energy levels to get closer together higher up the curve, as in Fig. 2.18, finally giving a continuum of levels.

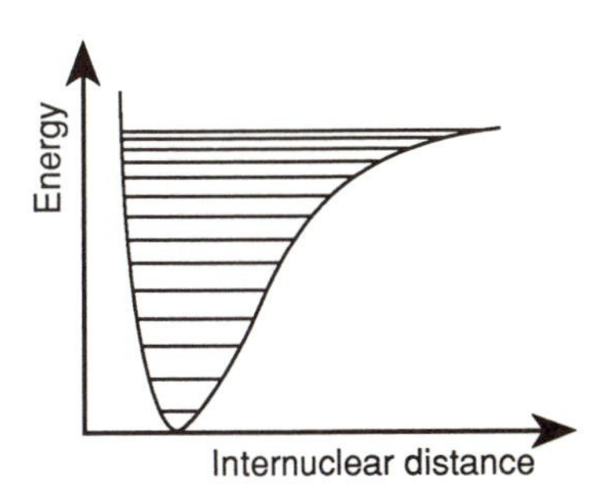

Fig. 2.18 Energy levels of an anharmonic oscillator.

An equation which will fit this behaviour well is:

$$E = (v + \tfrac{1}{2})\,h\nu - (v + \tfrac{1}{2})^2 x_e\,h\nu + (v + \tfrac{1}{2})^3 y_e\,h\nu$$

where x_e is a small anharmonic correction term, and further terms are increasingly small, and may usually be neglected.

For a harmonic oscillator, the vibration frequency is related to the force constant of the bond, k, by:

$$\nu = 1/2\pi\,(k/\mu)^{\frac{1}{2}}$$

where μ is the reduced mass of the system, $m_1 \cdot m_2/(m_1 + m_2)$. Thus from a knowledge of ν we can calculate the value of k, which is a rough measure of bond strength.

The potential energy curves of isotopically related molecules, such as H_2, HD, and D_2 are identical, as the forces holding the molecule together are independent of the nuclear masses. They therefore have the same force constant k, so we can predict the relationship between their vibrational energy levels. These are closer for the heavier isotopes, as shown in Fig. 2.19.

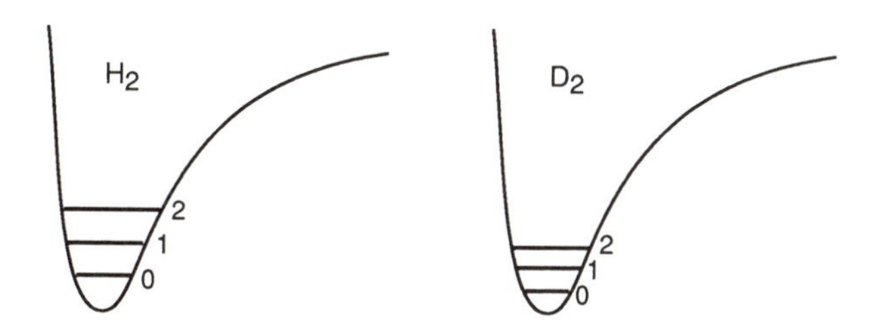

Fig. 2.19 Vibrational levels for different isotopic H_2 molecules.

One consequence of Fig. 2.19 is that the dissociation energies of H_2 and D_2 are slightly different. Although their potential energy curves are identical, D_2 has a smaller zero-point energy, and therefore more energy is needed to separate the atoms, starting from the lowest vibrational state; the distinction between the dissociation energies D_0 and D_e is shown in Fig. 2.20. This applies similarly to larger molecules containing C–H and C–D atoms; the C–H bonds are slightly easier to break. Reactions whose rate determining steps involve the breaking,

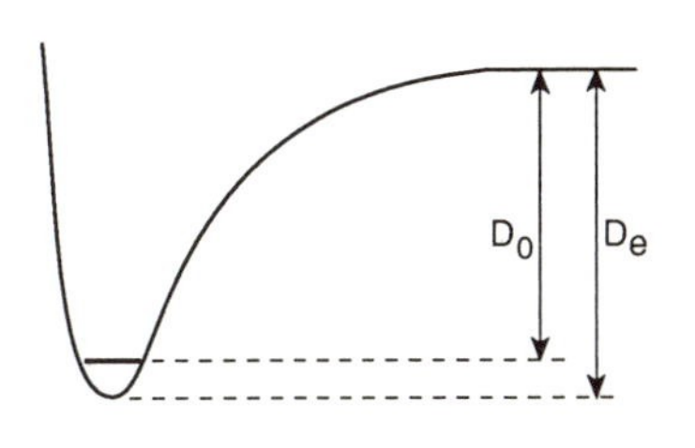

Fig. 2.20 The dissociation energies D_0 and D_e.

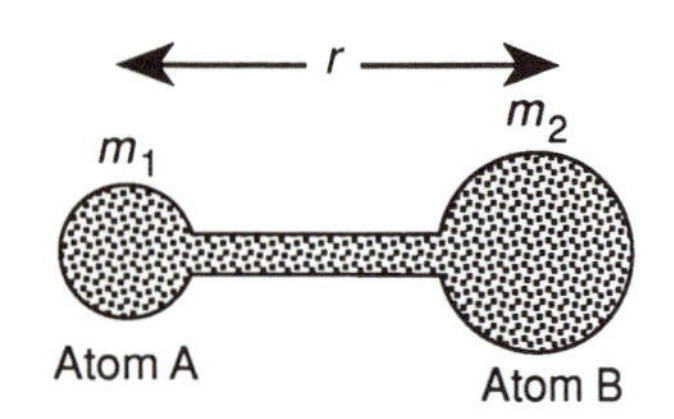

Fig. 2.21 The rigid rotor.

or partial breaking, of a C–H bond therefore proceed more slowly if the H atom is replaced by D; this effect, the kinetic isotope effect, is well known in organic chemistry, and gives important information on reaction mechanisms.

2.9 Rotational energy levels

Molecules can rotate as well as undergo vibrations; not surprisingly, we find that rotation is also quantized, with only certain definite energies allowed. The rate of vibration of a molecule is so rapid by comparison with the rate of rotation that we can consider a molecule to have an internuclear separation which is fixed at the value averaged over a complete vibration. The molecule thus becomes a rigid rotating body, as shown in Fig. 2.21.

The actual form of the rotational levels can be found by solving the Schrödinger equation for a rigid rotor, which leads to the energy levels and wave functions shown in Fig. 2.22. The levels are described by a quantum number J, and their energies are given by:

$$E_J = BJ(J+1)$$

where

$$B = h^2/8\pi^2\mu r^2$$

and μ is the reduced mass of the molecule:

$$\mu = m_1 \cdot m_2/(m_1 + m_2).$$

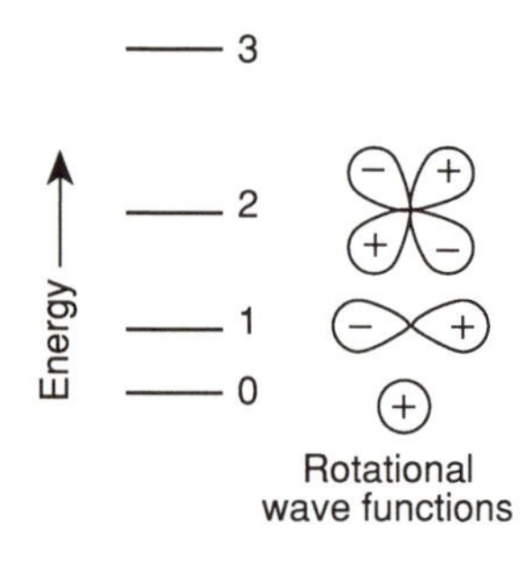

Fig. 2.22 Rotational energy levels and wave functions for a rigid rotor.

It is interesting to note that the rotational wave functions are the spherical harmonic functions which appeared in the solution to the Schrödinger equation for the hydrogen atom. The functions have a symmetry with respect to inversion at the origin which is alternatively symmetric (no change of sign) and antisymmetric (sign change).

The form of the energy levels of a rigid rotor can also be understood in terms of the relationship between the quantum number J and the angular momentum of the molecule. A full treatment shows that the angular momentum is quantized in units of $h/2\pi$; the exact relationship is:

$$P = (h/2\pi)\sqrt{J(J+1)}$$

This is exactly analogous to the role of the quantum number l in describing the orbital angular momentum of an atom. Now classically the angular momentum P of a rotor is equal to the moment of inertia I multiplied by the angular velocity ω; the moment of inertia is equal to the reduced mass multiplied by the bond length squared. The energy of a rotor is therefore given by:

$$\begin{aligned} E &= \tfrac{1}{2}I\omega^2 \\ &= \tfrac{1}{2}(I\omega)^2/I \\ &= h^2J(J+1)/8\pi^2 I \end{aligned}$$

$$= h^2J(J+1)/8\pi^2\mu r^2$$
$$= BJ(J+1),$$

as above.

Just as in the atomic case, there is a further quantum number M_J defining the component of J in a given direction; M_J can take any whole number value between $+J$ and $-J$. In the absence of an external field, states with differing values of M_J are all degenerate, but the fact that the number of degenerate states increases with J has consequences in determining the populations of different J levels. If the molecule is in an external field, then states with different M_J values are no longer degenerate; the splitting between such states in an electric field can be used to determine the dipole moment of the molecule.

In practice it is found that small deviations occur from the formula predicted for a rigid rotor; this is because as the molecule rotates, so it stretches slightly due to centrifugal distortion. This can be allowed for by introducing a correction term whose importance increases as J increases:

$$E_J = BJ(J+1) - DJ^2(J+1)^2.$$

Typically D is very much smaller than B; for HCl, $B = 10.35\ \mathrm{cm}^{-1}$ and $D = 0.0004\ \mathrm{cm}^{-1}$.

2.10 Rotational energy levels and nuclear spin

If we consider molecules which contain two equivalent nuclei, such as H_2 or O_2, or indeed larger molecules such as CO_2 or HCCH, then there is one further symmetry property which we need to take into account, namely the exchange of nuclei. This might seem a very small detail, but it has some major and unexpected consequences, and demonstrates the considerable power of quantum mechanics in describing molecular structure.

We have already seen in the simple case of the two electrons in a He atom that the wave function must be antisymmetric with respect to electron exchange. This prevents two electrons from having the same spin if they are in the same orbital, which is the generalisation known as the Pauli principle.

Electrons have spin $\frac{1}{2}$, and many nuclei also have half-integral spins; nuclear spin is described by the quantum number I, so for a ^{1}H nucleus $I = \frac{1}{2}$, and for a ^{35}Cl nucleus $I = \frac{3}{2}$. For particles with half-integral spin, the wave function must always be antisymmetric with respect to particle exchange. They are said to obey Fermi–Dirac statistics.

Other nuclei, such as ^{16}O and ^{12}C for which $I = 0$, and ^{2}H, for which $I = 1$, have integral spins, and these have wave functions which are symmetric to particle exchange. They are said to obey Bose–Einstein statistics.

Carbon dioxide

The simplest example of the effect of nuclear spin on molecular structure is carbon dioxide, where the two oxygen atoms are equivalent. We can see the effect of nuclear exchange on the total wave function of the molecule by considering its effect on the electronic, vibrational, and rotational wave functions separately, and then multiplying the results. The electronic wave function for CO_2 is ${}^1\Sigma_g^+$, and this is the case for the great majority of stable symmetric molecules. Nuclear exchange leaves the electronic wave function unchanged in this case. The vibrational wave function depends only on the magnitude of the bond length, and this is also unchanged on nuclear exchange. However the rotational wave functions are altered by nuclear exchange; when J is even, the wave function is symmetric, but when J is odd, the wave function is antisymmetric. The total molecular wave function is therefore symmetric to exchange of nuclei, as required by Bose–Einstein statistics, only if the rotational quantum number J is even. The remarkable conclusion from this analysis is that for the CO_2 molecule, J can only take the values, 0, 2, 4, etc. Even more remarkably, if we consider the molecule ${}^{16}OC^{18}O$, in which the oxygen nuclei are no longer equivalent, the restrictions above do not apply, and J may take the values 0, 1, 2, 3, etc.

It might be thought that the O_2 molecule would have been a more obvious example to choose to illustrate the effect of Bose–Einstein statistics on rotational states. Indeed the argument is very similar in this case, but because the electronic state is ${}^3\Sigma_g^-$, and antisymmetric with respect to exchange of nuclei, it is now only the rotational states with odd values of J which can exist.

Ortho- and para-hydrogen

We can now consider the effect of nuclear spins on the H_2 molecule, where the wave function must be antisymmetric with respect to nuclear exchange. As in the case of CO_2, the electronic and vibrational wave functions are symmetric with respect to nuclear exchange, and the rotational wave function is symmetric if J is even, and antisymmetric if J is odd.

The possible nuclear spin wave functions are more complex. Each nucleus has spin $\frac{1}{2}$, and we can write the two orientations as α and β. The possible nuclear spin wave functions are now:

$$\begin{array}{lll}
\alpha(1)\,\alpha(2) & T=1, & M_T=1 \\
\beta(1)\,\beta(2) & T=1, & M_T=-1 \\
\alpha(1)\,\beta(2)+\alpha(2)\,\beta(1) & T=1, & M_T=0 \\
\alpha(1)\,\beta(2)-\alpha(2)\,\beta(1) & T=0, & M_T=0
\end{array}$$

The first three wave functions are those associated with the total nuclear spin $T=1$ (the degeneracy $2T+1$ representing the number of possible orientations of T, each with its own value of M_T); the fourth represents the total nuclear spin $T=0$. The first three wave functions are symmetric with respect to nuclear exchange, and the fourth is antisymmetric.

For the whole wave function to be antisymmetric to nuclear exchange, it therefore follows that rotational levels with even values of J are associated with the total nuclear spin $T=1$, and odd values of J with total nuclear spin $T=0$. As the nuclear spin degeneracies are in the ratio 3:1, it is found that at high temperatures the populations of odd and even rotational levels are in the ratio 3:1.

Although molecules can change their rotational quantum numbers on collision, or by absorbing light, they cannot change their nuclear spins in these ways; if molecules with even values of J are separated from those with odd values, no interconversion occurs, even after long periods of time. Molecules with odd values of J are called ortho-hydrogen, and those with even values para-hydrogen; interconversion of the two forms only occurs when the nuclear spins can be reoriented, for example by dissociation and recombination, or by the magnetic field of a paramagnetic compound. If such interconversion takes place at very low temperatures, where only the lowest rotational level ($J = 0$) is populated, then only para-hydrogen is formed.

2.11 Excitation energies and populations

So far we have considered the patterns of electronic, vibrational, and rotational energy levels separately; now we must put them together, and compare the relative magnitudes of excitation energies.

The rotational energy levels are normally the most closely spaced; their separation depends on the value of B, whose value is determined mainly by the reduced mass μ. For molecules such as CO, B is about 1 cm^{-1}; for diatomics containing a hydrogen atom, the reduced mass is smaller, and the value of B is greater by a factor of about ten.

The unit cm^{-1}, the number of waves that will fit into 1 cm, is commonly used as a unit of energy; we can convert it readily into other more familiar energy units. Thus for HCl:

$$\begin{aligned}
B &= 10.35 && \text{cm}^{-1} \\
&= 10.35 \times c && \text{Hz} \\
&= 10.35 \times 3 \times 10^{10} && \text{Hz} \\
&= 3.1 \times 10^{11} && \text{Hz} \\
&= 3.1 \times 10^{11} \times h && \text{J} \\
&= 3.1 \times 10^{11} \times 6.6 \times 10^{-34} && \text{J} \\
&= 2.1 \times 10^{-22} && \text{J.}
\end{aligned}$$

We can utilize this value of B in Joules to calculate the bond length of HCl, using the equation given in Section 2.9; if the reduced mass is calculated in kg, then the bond length comes out in metres. The value of B can also be converted into the more familiar J mol^{-1}:

$$\begin{aligned}
B &= 2.1 \times 10^{-22} \times 6 \times 10^{23} && \text{J mol}^{-1} \\
&= 130 && \text{J mol}^{-1} \\
&= 0.13 && \text{kJ mol}^{-1}.
\end{aligned}$$

Vibrational frequencies are often also expressed in the units cm^{-1}, although properly frequencies should be measured in Hz; as above, the wavenumber in cm^{-1} needs to be multiplied by c, the speed of light, in cm s^{-1} to produce the frequency in Hz. Vibrational frequencies are typically of the order of 10^{13} Hz, corresponding to roughly 1000 cm^{-1}, or about 10 kJ mol^{-1} between vibrational energy levels.

Electronic excitation energies are typically larger than vibrational energies by another factor of 10 to 100, being of the order of 100 kJ mol^{-1}.

The average kinetic energy of molecules is approximately RT, where R is the gas constant, 8.3 J mol^{-1} K^{-1}; at room temperature therefore, molecules have an average kinetic energy of a few kJ mol^{-1}. The consequence of this is that at room temperature, molecules normally occupy only the most stable electronic and vibrational states, as the energy required to excite them to higher states is not available in ordinary collisions. On the other hand, the energy required for rotational excitation is less than that available in typical collisions, and so a number of rotational energy levels will be significantly populated. The precise calculation of rotational populations needs the $(2J + 1)$ degeneracy of higher states to be taken into account.

3 Polyatomic molecules

3.1 Orbitals in polyatomic molecules

The principles that we have used to understand the electronic structures of diatomic molecules can be carried across to larger molecules. Molecular orbitals can be constructed from linear combinations of atomic orbitals; the total electronic wave function is then antisymmetrized to take account of the Pauli principle. As before, we can construct a molecular-orbital diagram which is qualitatively correct by using just the atomic orbitals which are occupied in the separate atoms.

When we considered diatomic molecules, we saw that their orbitals could be classified by their symmetry properties. Thus orbitals were described σ if they had cylindrical symmetry, and π if they were of opposite sign above and below the internuclear axis. Similarly in the case of homonuclear diatomics, such as O_2, the orbitals were classified as u or g, depending on whether the wave function changed sign on inversion at the centre of the molecule. These classifications are useful because they help us to identify which atomic orbitals can be used to build molecular orbitals; orbitals of different symmetries cannot overlap forming bonding or anti-bonding orbitals. In this way an s orbital on one atom can bond with a p_z orbital on an adjacent atom, but not with a p_x or a p_y orbital.

In the case of diatomic molecules, simple diagrams can show the symmetries of orbitals; for polyatomic molecules, the symmetries may be harder to visualize. Descriptions of molecular symmetry can be formalized using group theory, which turns out to be particularly useful in describing the structures of more complex molecules. A full treatment of group theory is not needed at this stage, but we can see how it can be applied by considering a simple example, H_2O; we shall also see the origin of the notation often used to describe molecular orbitals and electronic states.

3.2 Water, H_2O

Figure 3.1 shows the water molecule, with the axes labelled. The molecule has two symmetry planes, the yz- and the xz- planes, and a two-fold rotation axis about the z-axis. This means, for example, that the electron density must be unchanged if we rotate the molecule through 180° about the z-axis; it therefore follows that this operation must either leave the wave function unchanged, or simply change its sign, as the electron density depends on the square of the wave function. We can therefore classify orbitals according to how they behave under each symmetry operation.

The notation used in this case is as follows:

(a) Orbitals of types a_1 and a_2 are unchanged on rotation.

(b) Orbitals of types b_1 and b_2 change sign on rotation.

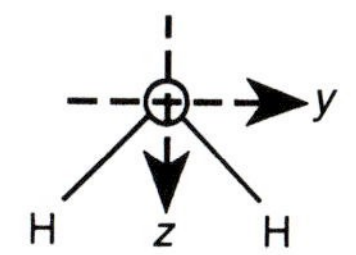

Fig. 3.1 Axes in the structure of the water molecule.

(c) Orbitals of types a_1 and b_2 are unchanged on reflection in the yz-plane.

(d) Orbitals of types a_2 and b_1 change sign on reflection in the yz-plane.

We can now classify the atomic orbitals which we shall use to construct the molecular orbitals for water. On the oxygen atom, the 1s, 2s, and $2p_z$ orbitals all have a_1 symmetry; the $2p_x$ has b_1 symmetry; and the $2p_y$ orbital has b_2 symmetry. For the hydrogen atoms, the situation is a little more complex; if we call the 1s orbitals on the atoms h_1 and h_2 respectively, then it is easy to show that the combination $(h_1 + h_2)$ has a_1 symmetry, and the combination $(h_1 - h_2)$ has b_2 symmetry.

We can now construct a molecular-orbital diagram for the water molecule. The two most stable orbitals are largely unchanged on bonding, and resemble closely the oxygen 1s and 2s atomic orbitals; they have symmetry a_1. The oxygen $2p_x$ orbital is not involved in bonding, as there are no other orbitals of the same symmetry; it alone changes sign on reflection in the yz-plane. It can therefore be described as a non-bonding orbital. There are two orbitals of symmetry a_1, the $2p_z$ orbital, and the $(h_1 + h_2)$ combination; as these are of comparable energies, they combine to form a bonding orbital and an anti-bonding orbital, each with symmetry a_1. Similarly the $2p_y$ orbital and the $(h_1 - h_2)$ combination combine to form a bonding and an anti-bonding orbital, each of symmetry b_2.

The molecular orbitals of water, in order of increasing energy, are therefore:

$1a_1$	1s (O)	non-bonding;
$2a_1$	2s (O)	non-bonding;
$3a_1$	2p (O) and 1s (H)	bonding;
$1b_2$	2p (O) and 1s (H)	bonding;
$1b_1$	2p (O)	non-bonding;
$2b_2$	2p (O) and 1s (H)	anti-bonding;
$4a_1$	2p (O) and 1s (H)	anti-bonding;

The ten electrons occupy the five most stable orbitals, giving four electrons in bonding orbitals, and none in anti-bonding orbitals; this corresponds to the classical description of water as containing two single bonds.

There is one feature of this description of the water molecule which is at first sight rather surprising. The bonding orbitals $3a_1$ and $1b_2$ are both spread over all three atoms of the molecule; there is no direct relationship between a particular bonding orbital and a particular covalent bond. This is unexpected because there is much chemical evidence about bond energies which suggests that bonding electrons can be seen as localized between pairs of adjacent atoms, rather than spread over the whole molecule.

This discrepancy can be resolved, at least for molecules which contain closed electron shells. The electronic wave function is in fact a determinant, and it can be shown that the value of a determinant is unchanged if any multiple of one row is added to another. This means that we may take linear combinations of the orbitals that we have produced above without altering the overall electron density at all. If we take the combination $(3a_1 + 1b_2)$ we shall produce a new localized orbital, in which the electrons are shared between the O atom and one of the H atoms; the combination ($3a_1 - 1b_2$) similarly produces an orbital localized between the O atom and the other H atom. This process does not involve us altering the overall wave function at all, merely changing the way in which we perceive it. We shall see it again when we discuss the phenomenon of hybridization.

3.3 Molecular shape

So far we have described the water molecule as being formed by the overlap of the 1s orbitals on the H atoms with the $2p_y$ and $2p_z$ on the O atom; as the 2p orbitals are at right angles to each other, the bond angle in H_2O should be 90°. In the same way we can produce a description of the NH_3 molecule in which the 1s orbitals on the H atoms overlap with the three 2p orbitals, again producing bond angles of 90°.

This model has correctly identified that water is not linear, and ammonia is not planar, but its predicted bond angles are a little too low; the experimental values are 104° and 107°. The reason for this is that we have so far neglected any contribution to bonding from the 2s orbital. This orbital is more stable than the 2p orbital, but it is of the right symmetry (a_1 in H_2O). We may therefore expect that the $2a_1$ orbital will be mainly, but not entirely, 2s, and the $3a_1$ orbital will acquire some 2s character. It can be shown that the effect of mixing in some of the 2s orbital is to widen the bond angle, and this brings our prediction more closely in line with experiment.

There are two ways of thinking about the involvement of the 2s orbitals in bonding in water. One is to begin by omitting the 2s orbitals, as we did in the last section, and then modifying our orbitals by mixing in the 2s orbital. An alternative, and equally valid, method is to begin by mixing together the 2s and 2p orbitals, and then allowing the H 1s orbitals to overlap forming bonding and anti-bonding orbitals. This mixing together of 2s and 2p orbitals is often called *hybridization*. As we saw in the last section, the mixing of orbitals is a perfectly proper way of considering orbital shape; it leaves the total electron density unchanged, but it allows us to visualize the different components more clearly.

3.4 sp^3 hybridization in carbon compounds

The idea of hybridization is most familiar in organic chemistry. Carbon has the atomic structure:

$$1s^2\, 2s^2\, 2p_x\, 2p_y$$

Although the 2s orbital is fully occupied in the atom itself, it is energetically worthwhile to use this orbital in covalent bond formation. One way of expressing this is to note that the energy needed to promote a 2s electron to the 2p orbital is about 436 kJ mol^{-1}, while the formation of two extra C–H bonds is exothermic to the extent of nearly 900 kJ mol^{-1}. We can therefore consider ourselves starting from an excited carbon atom with the configuration:

$$1s^2\, 2s\, 2p_x\, 2p_y\, 2p_z.$$

Fig. 3.2 sp^3 hybrid orbitals.

This is however only a convenient way of imagining the energetics of bond formation; the carbon atom does not actually need to pass through this state.

We may now mix together, or hybridize, the four orbitals, one 2s and three 2p, to form four new orbitals, called sp^3 hybrids. These orbitals turn out to be pointing towards the corners of a regular tetrahedron, as shown in Fig. 3.2. Each of these orbitals can now overlap very effectively with a H 1s orbital, leading to four single covalent bonds.

It is important to remember that hybridization is not really an effect, but rather a different way of looking at a wave function. Our wave functions are really determinants, so that for the excited carbon atom we could write:

$$\Psi = \begin{vmatrix} s(1) & p_x(1) & p_y(1) & p_z(1) \\ s(2) & p_x(2) & p_y(2) & p_z(2) \\ s(3) & p_x(3) & p_y(3) & p_z(3) \\ s(4) & p_x(4) & p_y(4) & p_z(4). \end{vmatrix}$$

The hybridized orbitals are of the form:

$$\begin{aligned} a &= s + p_x + p_y + p_z \\ b &= s - p_x - p_y + p_z \\ c &= s + p_x - p_y - p_z \\ d &= s - p_x + p_y - p_z \end{aligned}$$

and so the hybridized wave function is

$$\Psi' = \begin{vmatrix} a(1) & b(1) & c(1) & d(1) \\ a(2) & b(2) & c(2) & d(2) \\ a(3) & b(3) & c(3) & d(3) \\ a(4) & b(4) & c(4) & d(4). \end{vmatrix}$$

Multiplying out the two determinants in full shows that the hybridized and unhybridized forms are exactly the same.

3.5 Other hybridizations

Classical organic chemistry has shown that carbon can also form double and triple bonds. To describe a molecule such as methane we visualized sp^3 hybrids on the carbon atom. For double-bonded molecules such as C_2H_4 we combine one s and two p orbitals to give planar sp^2 hybrids, leaving one p orbital unchanged, at right angles to the hybrids (Fig. 3.3).

Four of the hybrid orbitals overlap with the hydrogen atom 1s orbitals forming covalent bonds; in each case these are cylindrically symmetrical about the bond axis, and are called σ bonds. The two remaining hybrids, one on each C atom, overlap forming the C–C σ bond. The two unused p orbitals now overlap sideways; this forms an orbital whose sign is positive above the molecular plane, and negative below it, and is thus a π bond.

If the electron densities and nuclear charges were all identical, the bond angles in sp^2 hybridization would all be 120°; in C_2H_4, differences in electron repulsions mean that the HCH angle is about 117°.

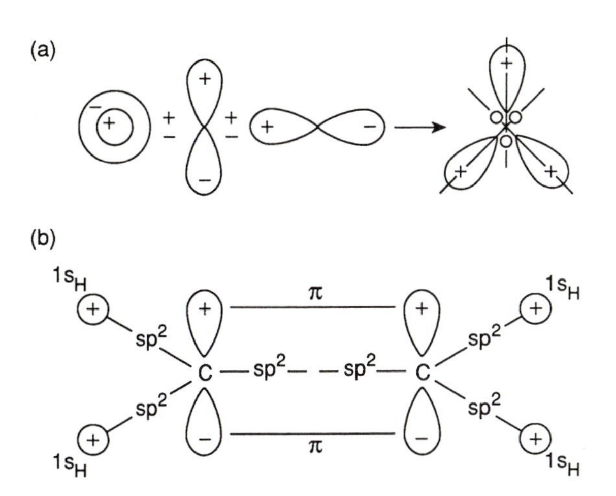

Fig. 3.3 (a) Formation of sp^2 hybrids. (b) Bonding in ethene.

For triple-bonded molecules, each carbon atom can be visualized as forming two sp hybrids, leaving two p orbitals unhybridized. In C_2H_2 the sp hybrids then form C–H and C–C σ bonds, and the unhybridized p orbitals overlap sideways again, forming two π orbitals. This is shown in Fig. 3.4; the second π orbital is above and below the plane of the paper, so it is not shown.

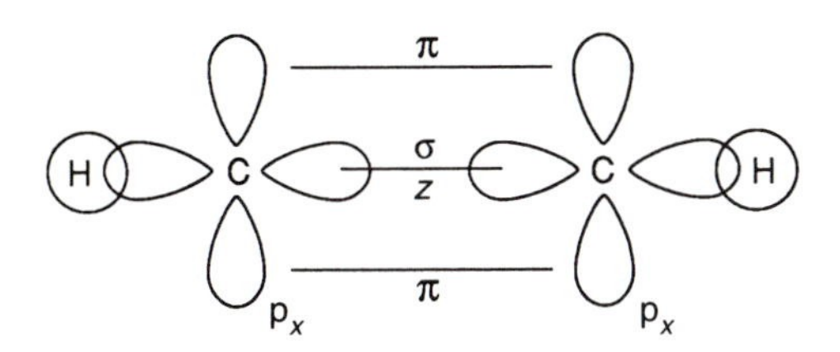

Fig. 3.4 Bonding in ethyne.

The central C atom in the molecule $H_2C{=}C{=}CH_2$ is sp hybridized: the two π orbitals, corresponding to the two double bonds, are therefore at right angles to each other, which explains why the molecule is not planar.

Hybridization can use d orbitals as well as s and p orbitals; examples are found in transition-metal complexes, as well as non-metals with more than eight electrons in their outer shells. Table 3.1 summarizes possible hybridizations, and the resulting geometric arrangements.

3.6 Delocalization in benzene

We saw that in ethene a π bond is formed by the overlap of two p orbitals not involved in the formation of σ bonds. A similar situation arises in benzene, as

Table 3.1 Summary of hybridizations

Coordination no.	Atomic orbitals	Geometry
2	sp	linear
	pd	linear
	sd	bent
3	sp^2	trigonal plane
	dp^2	trigonal plane
	ds^2	trigonal plane
	d^2p	trigonal pyramid
4	sp^3	tetrahedral
	sd^3	tetrahedral
	dsp^2	square plane
5	dsp^3	bipyramid
	d^3sp	bipyramid
	d^4s	tetragonal pyramid
6	d^2sp^3	octahedral
	d^4sp	trigonal prism

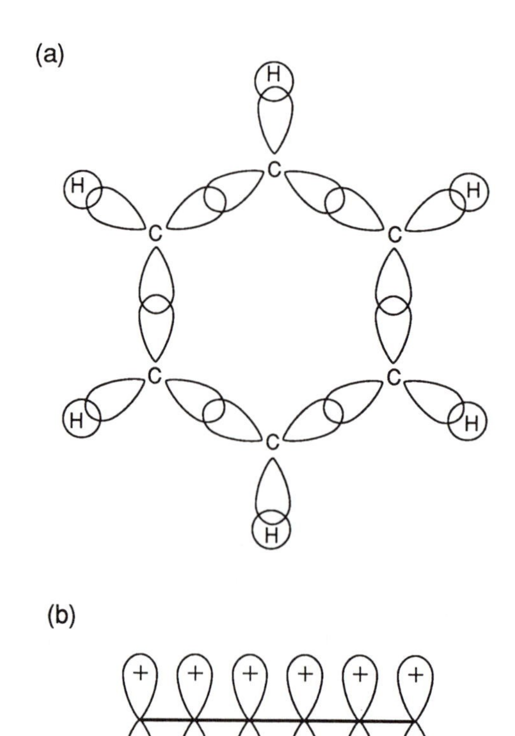

Fig. 3.5 Molecular orbitals in benzene: (a) xy-plane; (b) z direction.

shown in Fig. 3.5. Each of the six carbon atoms is sp^2 hybridized, and forms three σ bonds, one to each adjacent carbon atom, and one to hydrogen. This leaves six electrons to occupy orbitals formed by the overlap of the six unused 2p orbitals, which are at right angles to the plane of the molecule.

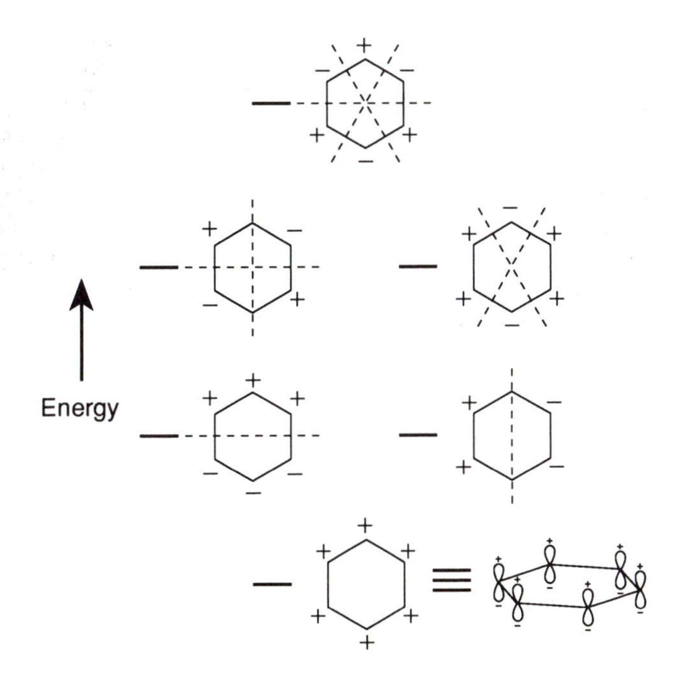

Fig. 3.6 Schematic representation of the π-molecular orbitals of benzene.

The resulting molecular orbitals can be constructed on symmetry grounds, and are shown in Fig. 3.6. As we begin with six atomic orbitals, there are six resulting molecular orbitals, of which three are bonding and three are anti-bonding. The most stable molecular orbital has no nodes at all (except in the molecular plane) and the orbitals become less stable as the number of nodes increases. The electrons occupy the three most stable orbitals, the three bonding orbitals, as shown in Fig. 3.7. Detailed calculations show that the total bonding energy exceeds that expected for three separate double bonds, which explains the high stability of the benzene ring.

A similar analysis for cyclobutadiene (C_4H_4) shows that there are four π molecular orbitals formed, one bonding, one anti-bonding, and two non-bonding. The total bonding energy is now much less than in benzene, and indeed no better than would be expected for separate double bonds; this agrees with observations that cyclobutadiene has no aromatic character. Aromaticity is only observed when the number of π electrons is given by the formula $(4n+2)$.

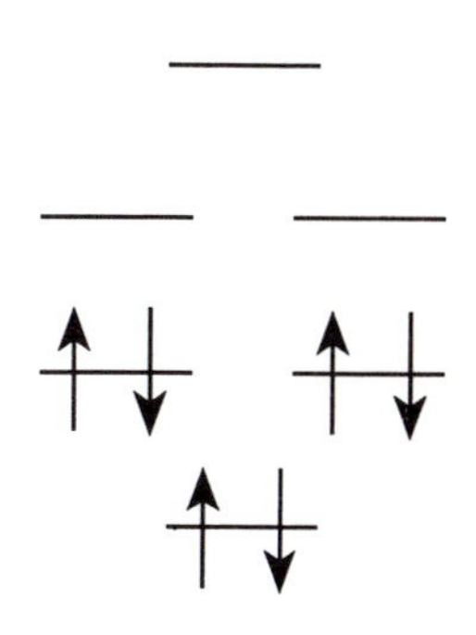

Fig. 3.7 The occupied molecular orbitals of benzene.

3.7 Excited states in organic compounds

Now that we have seen how the ground states of simple organic molecules can be described, we can go on to see how their excited states arise. One observation helps to simplify our description considerably: chemically similar molecules often have almost identical excitation energies. Thus all carbonyl compounds have an excited electronic state about 400 kJ mol^{-1} above the ground state. The excitation can therefore be seen as localized almost entirely in the C=O bond. Groups which give rise to characteristic excitation energies are called chromophores.

Alkanes require considerable energy to excite their electrons from a bonding σ orbital to an anti-bonding σ^* orbital. These excitations are relatively difficult to observe spectroscopically, as they require photons in the vacuum ultra violet, and they are of little practical significance. Cycloalkanes such as cyclohexane generally behave similarly, but cyclopropane is an exception; excitation to an anti-bonding σ^* orbital requires significantly less energy, as the bonding orbitals are destabilized by ring strain.

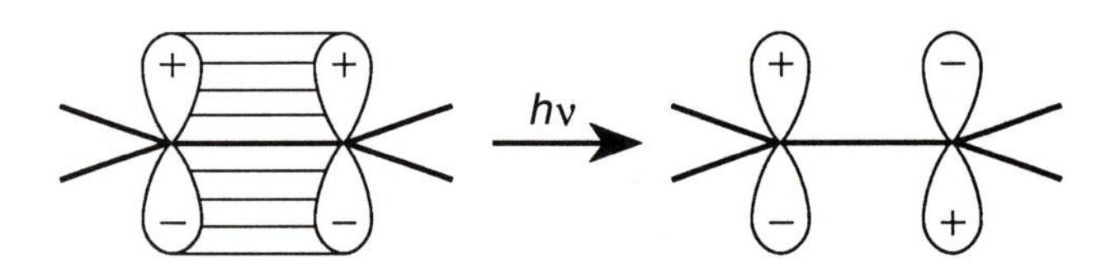

Fig. 3.8 The (π^*, π) transition in ethene.

Alkenes can be excited by absorption of U.V. photons of wavelength about 190 nm; a bonding electron in a π orbital is thereby promoted to an anti-bonding π^* orbital, as shown in Fig. 3.8. The variation of this excitation energy when the H atoms in C_2H_4 are replaced by other atoms has been extensively studied, and precise measurement of the excitation energy can give structural information. In the excited state the π-bond has been broken, and the barrier to the rotation about the double bond is removed. The absorption of U.V. radiation can therefore be followed by rotation, and subsequent reforming of the bond can lead to a different isomer being produced. The excitation of electrons from a σ orbital, or to a σ^* orbital, requires much more energy, and is not usually observed.

If a molecule contains two double bonds which are well separated from each other, then each behaves as described in the paragraph above. However if the double bonds are adjacent, as in CH_2=CH–CH=CH_2, their π orbitals overlap, and the system is said to be conjugated. The energy required to excite an electron from a π to a π^* orbital is reduced. This can be understood by considering the molecular orbitals formed from the overlap of four p orbitals, as shown in Fig. 3.9. Four molecular orbitals are formed, two bonding and two anti-bonding; as we have seen before, the orbital with the largest number of nodes has the highest energy. The four electrons occupy the bonding orbitals ϕ_1 and ϕ_2. Fig. 3.10 gives a simple molecular-orbital diagram, which shows how

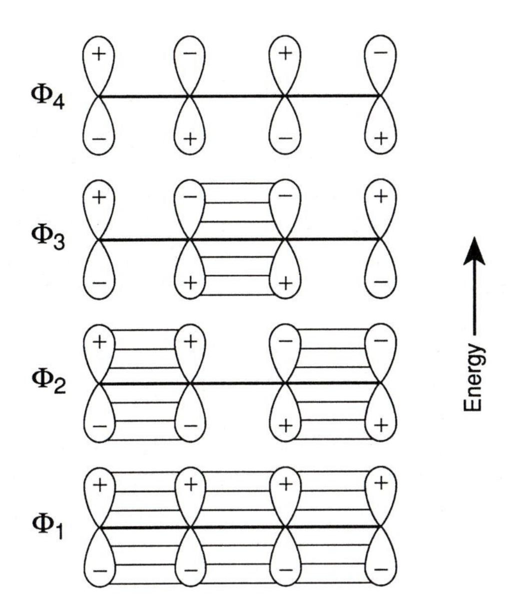

Fig. 3.9 π-orbitals in butadiene.

the excitation energy decreases on conjugation. If the conjugation is lengthened to include more double bonds then the excitation energy falls further, and eventually corresponds to photons in the visible region of the spectrum. Many natural products have long conjugated structures, and this π to π^* excitation is the reason why carotenes, and hence carrots, are coloured.

The effect of conjugation can also be used to obtain conformational information on biphenyls. The excitation energies of C_6H_5–C_6H_5 and of $CH_3C_6H_4$–$C_6H_4CH_3$ are very similar to each other, provided that the methyl groups occupy the 3- or 4- positions in the rings. However if the two methyl groups occupy the 2-positions, the excitation energy increases markedly. This is because biphenyls are normally planar, with the π electron systems conjugated. If the methyl groups are in the 2-position, steric hindrance twists one ring relative to the other, and conjugation is no longer possible.

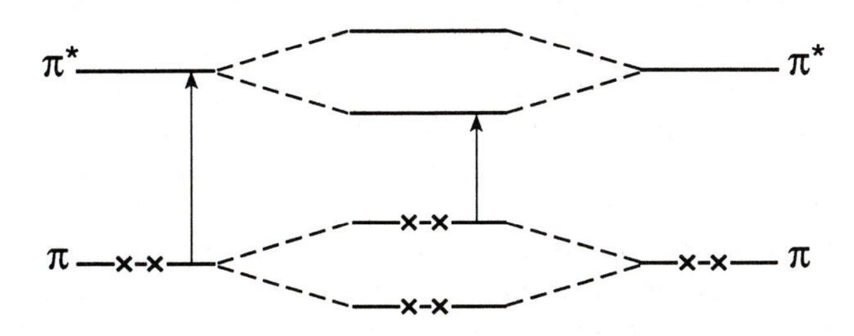

Fig. 3.10 The interaction of two π-orbitals in butadiene.

Aldehydes and ketones have excited states corresponding to the promotion of a non-bonding electron on the oxygen atom to an anti-bonding π^* orbital. This excitation can be achieved by a photon of wavelength of about 290 nm; it is represented in Fig. 3.11. If the C=O bond is conjugated with a C=C bond, then the excitation energy is decreased. There are also excited states known in which a non-bonding electron is excited to an anti-bonding σ^* orbital, and with a π electron excited to an anti-bonding π^* orbital; these have higher excitation energies. Similar excited states are known in related inorganic ions such as carbonate and nitrate.

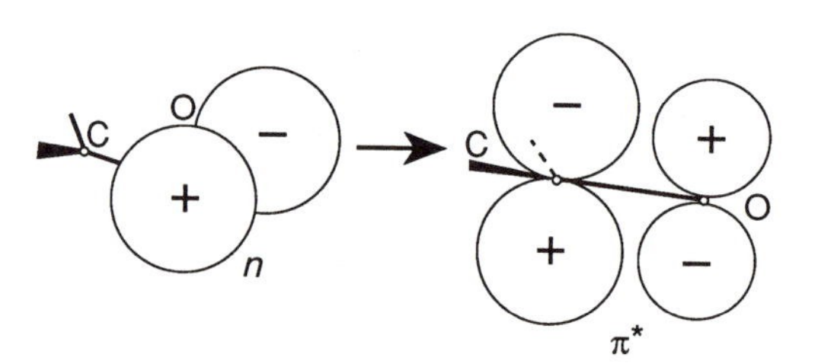

Fig. 3.11 The (π^*, n) transition in H_2CO.

We have been concerned in this book mainly with the properties of molecules in the gas phase; many experiments on organic molecules are, however, performed in solution. The effects of non-polar solvents are often very small, and excitation energies are little changed from the gas phase. The use of polar solvents can produce more significant changes. In carbonyl compounds the ground state is more polar than the excited state formed by the excitation of an electron from n to π^*, and so interacts more strongly with the solvent. The excitation energy is therefore increased. This sort of argument has been helpful in assigning excited states, although there are several well known exceptions.

3.8 Walsh diagrams

The shapes of small molecules are determined by three factors, the attraction between electrons and nuclei, the repulsions between electrons, and the repulsions between nuclei. Although quite sophisticated calculations are needed to determine the exact variation of each of these, some simple approximations can often result in reliable predictions. The best known of these is the electron pair repulsion theory, which states that the shape of a molecule can be predicted by minimizing the repulsion between electron pairs, bonding or non-bonding, and neglecting all other factors. In this way water is predicted to be bent, methane to be tetrahedral, and boron trifluoride to be planar.

Another approach to molecular shape is offered by Walsh diagrams, which show how the energies of molecular orbitals vary with geometry. Although the molecular orbital energy is not equal to the total energy of the molecule, they can be used to predict the shapes of both ground and excited states of small molecules.

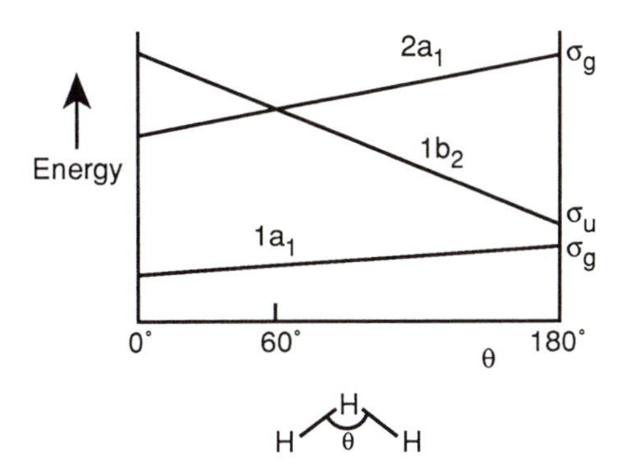

Fig. 3.12 Schematic Walsh diagram for H_3.

The Walsh diagram for H_3 shown in Fig. 3.12. This molecule has orbitals of symmetry a_1, b_2, and a_1; as the molecule becomes linear, these become σ_g, σ_u, and σ_g. Now the ion H_3^+ contains two electrons, and so the electron configuration will be $1a_1^2$. The electron energy therefore becomes more favourable as the bond angle becomes smaller. Now nuclear repulsion will also be an important factor as the bond angle decreases, but it is clear that the H_3^+ ion will be bent, not linear. The H_3 molecule will contain one more electron; the most stable arrangement will put this in the $1b_2$ orbital. The molecule is therefore predicted to be linear, not bent. Experimentally it has been shown that both of these conclusions are correct.

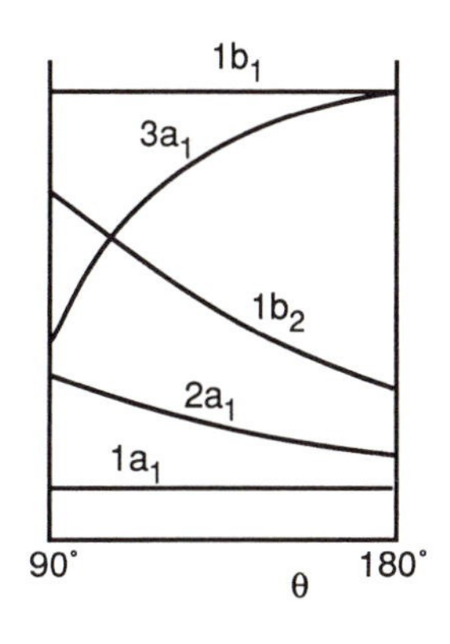

Fig. 3.13 Schematic Walsh diagram for AH_2 molecules.

Figure 3.13 shows the Walsh diagram for species of the sort AH_2. In this case it can be seen that six electron species such as BeH_2 and BH_2^+ should be linear, but those with more electrons, such as BH_2, CH_2, and H_2O should be bent. The experimental values are: BeH_2 180°, BH_2 131°, CH_2 103°, H_2O 104°. The diagram can also be applied to excited states: thus for BH_2:

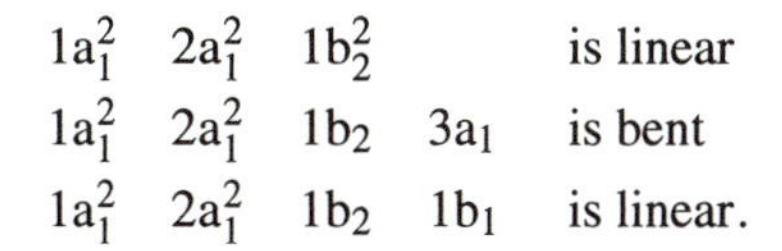

$1a_1^2$	$2a_1^2$	$1b_2^2$		is linear
$1a_1^2$	$2a_1^2$	$1b_2$	$3a_1$	is bent
$1a_1^2$	$2a_1^2$	$1b_2$	$1b_1$	is linear.

3.9 Transition metal complexes

Transition metal complexes almost always contain the transition metal in a positive oxidation state; in these circumstances we may neglect the s and p orbitals, and consider only metal d orbitals. The simplest model, the crystal field model, then describes the complex as a positively charged metal ion, surrounded by ligands which are thought of as point negative charges.

In the absence of any ligands, the five d orbitals are all of the same energy, that is degenerate. The effect of introducing six ligands arranged in a regular octahedron, as in Fig. 3.14, is to lift this degeneracy. The orbitals are split into two degenerate groups: three are labelled t_{2g}, and these are more stable than

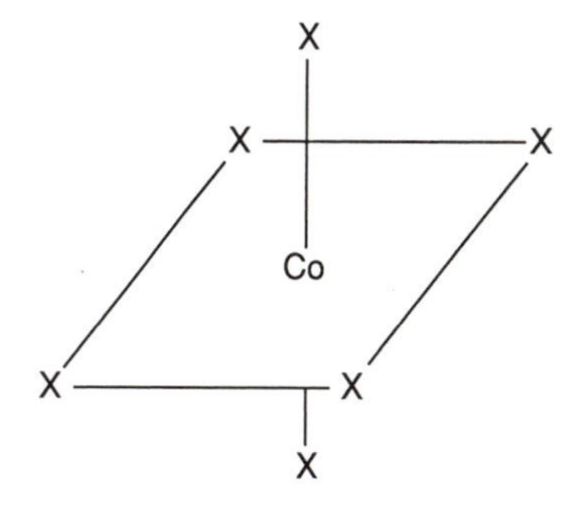

Fig. 3.14 An octahedral complex.

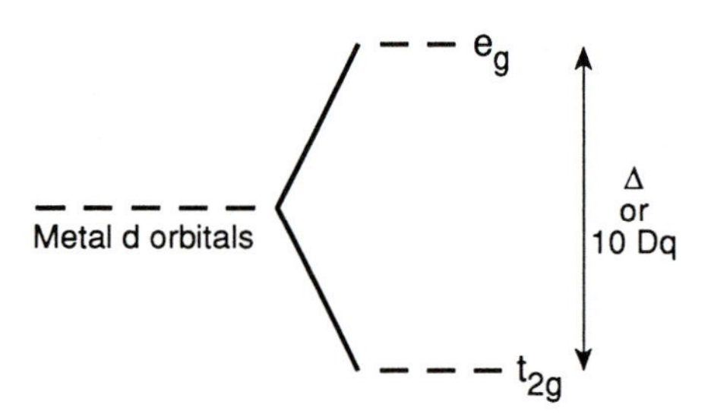

Fig. 3.15 Crystal field splitting of metal d orbitals in an octahedral field.

the other two, which are labelled e_g. Fig. 3.15 shows the energy level diagram for the metal ion. The energy gap between the t_{2g} and e_g orbitals is known as 10 Dq, for historical reasons.

We can now apply the aufbau principle to our energy level diagram. If we have a complex with the configuration d^1, such as Ti(III), then that electron will occupy a t_{2g} orbital. For complexes with two d electrons, two separate t_{2g} orbitals will be occupied, so as to minimize electron repulsion; the electrons will have parallel spins, in accordance with Hund's rule.

A more difficult situation arises in complexes with four d electrons, such as Mn^{3+}. The fourth electron may now go into a t_{2g} orbital, suffering relatively high repulsion from the other electron in that orbital, or it may occupy an e_g orbital, with its less favourable interaction with the ligands. These alternatives are shown in Fig. 3.16, and are known as low-spin and high-spin respectively. Which arrangements actually occurs depends on the magnitude of the crystal field splitting 10 Dq; it has an important effect on the magnetic and optical properties of the complex. Similar situations arise with ions containing 5, 6, or 7 d electrons.

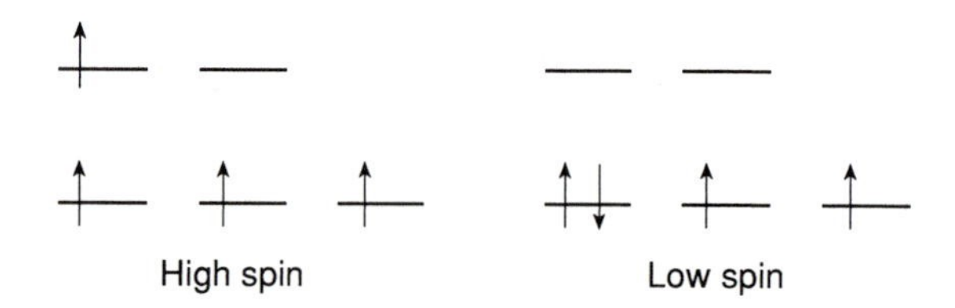

Fig. 3.16 Low-spin and high-spin configurations.

The magnitude of the crystal field splitting 10 Dq depends on the natures of the ligand and the transition metal. Ligands may be arranged in order of the splittings they produce; for some common ligands the order is:

$$I^- < Br^- < Cl^- < F^- < H_20 < NH_3 < NH_2CH_2CH_2NH_2 < CN^-$$

where I^- produces the smallest splittings. The crystal field theory treats all splittings as being purely electrostatic, but this does not explain all the observed data quantitatively. More satisfactory agreement is obtained using a molecular-orbital model which includes ligand orbitals as well as the d orbitals, therefore allowing some degree of covalency between the ligands and the metal.

One further feature of the shape of transition metal complexes is the distortion which occurs because of the Jahn–Teller theorem. This theorem states that 'any non-linear molecule in a degenerate electronic state will be unstable, and undergo a distortion which will lower its symmetry and split the degenerate state'. This theorem is most commonly encountered in transition metal complexes; thus the ion CoF_6^{3-} is not a regular octahedron, but is elongated along one axis.

3.10 Excited states and transition metals

It is well known that many transition metal ions are coloured. Table 3.2 gives the colours of some simple ions in the first transition series. In these ions the colour is caused by an incomplete d shell; only ions with an empty d shell or a full d shell are colourless, and the remainder are all coloured.

Table 3.2 Colours of some simple ions in the first transition series

K^+, Ca^{2+}, Sc^{3+}	d^0	colourless	Fe^{2+}	d^6	green
Ti^{3+}	d^1	pink	Co^{2+}	d^7	pink
V^{3+}	d^2	blue	Ni^{2+}	d^8	green
Cr^{3+}	d^3	green	Cu^{2+}	d^9	blue
Fe^{3+}	d^5	yellow	Zn^{2+}	d^{10}	colourless

The colour arises from the fact that transition metal ions have relatively low-lying excited states. The ion $Ti(H_2O)_6^{3+}$ has an excited state about 20 400 cm^{-1} above the ground state; this energy can be supplied by a photon of wavelength 490 nm, which is in the visible region of the spectrum. Figure 3.17 shows how this excited state can be understood in terms of crystal field theory; the ion contains a single d electron, which occupies not a t_{2g} orbital, but a less stable e_g orbital. This sort of excited state will always exist, as long as the d shell is neither completely full nor completely empty.

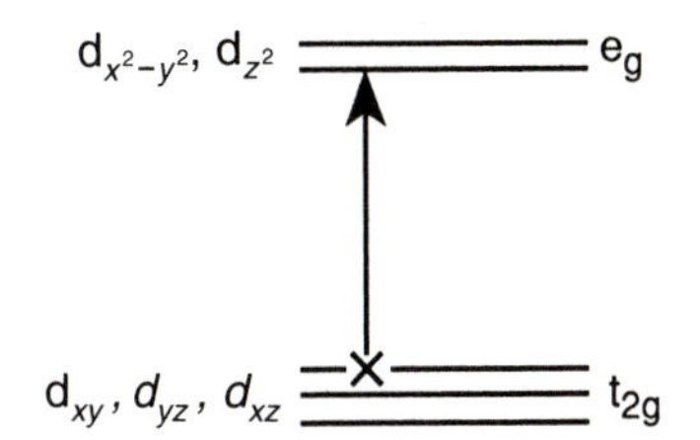

Fig. 3.17 Absorption of light by $Ti(H_2O)_6^{3+}$.

A similar argument can be applied to the excited states of tetrahedral complexes; again the ligands split the d orbitals into two groups, t_{2g} and e_g, but now the e_g orbitals are the more stable. The energy difference is still similar to that in the octahedral case, and so electrons can be excited from one d orbital to another by visible photons.

Although the colours of many transition metal ions can be explained in this way, the explanation does not apply in every case. The ions VO_4^{3-}, CrO_4^{2-}, and MnO_4^- are all coloured, even though their oxidation numbers correspond to the metal ion having no d electrons at all. In these cases, a different type of excited state gives rise to their colour. Here an electron in a ligand orbital is excited to an unoccupied d orbital on the metal ion. The excitation energy depends on the energy of the vacant d orbital; for the series VO_4^{3-}, CrO_4^{2-}, and MnO_4^-, the excitation energy drops steadily as the d orbital becomes more stable, that is as the ion becomes more oxidizing. These excitations are called charge-transfer transitions.

Similar excited states have been observed in many other transition-metal complexes, including those with partially filled d-shells; the excitation energies are often rather higher than in the examples above, and correspond to photons in the ultra violet region of the spectrum. Charge-transfer transitions are also known in weakly bonded complexes. Thus iodine is purple in the gas phase, and in solvents such as CCl_4; however in benzene and ether it appears brown. In the latter cases it forms a weak complex with the solvent, and the brown colour is due to the excitation of an electron in the solvent molecule to an anti-bonding orbital in the I_2 molecule.

3.11 Rydberg states

Most of the excited states that we have considered so far have involved the promotion of a single valence electron to a low-lying excited orbital; states involving more extensive excitations are relatively uncommon, and frequently unstable.

There is one type of highly excited state that has been recognized in both diatomic and polyatomic molecules, where an electron is promoted to an orbital whose radius is large compared with the size of the rest of the molecule. Under these circumstances the electron experiences an electric field which is roughly equal to that of a single positive charge at the centre of the molecule. The situation is now reminiscent of the excitation of a hydrogen atom. The excited orbitals are very diffuse, and have energies which follow the atomic pattern, fitting approximately the formula:

$$E = -R/n^2$$

where R is the Rydberg constant. These electronic levels are therefore called Rydberg states.

Since the observed energy levels can be fitted to such a simple formula, it is possible to extrapolate the energy levels to the point where the electron is no longer bound to the molecular framework. This provides a measure of the molecular ionization energy, in exactly the same way that extrapolating to series limits measures atomic ionization energies. This was originally an important method of measuring molecular ionization energies, but has now been largely replaced by photoelectron spectroscopy.

3.12 Photoelectron spectroscopy

Photoelectron spectroscopy, and other related techniques, have grown considerably in importance recently; the photoelectron experiment essentially measures molecular ionisation energies. Studies of the binding energies of individual electrons have provided information on inductive effects, delocalization of electrons, p$\rightarrow$ dπ bonding and molecular geometry, as well as finding analytical applications. Details of these uses of photoelectron spectroscopy can be found elsewhere; we are concerned here with the energy levels involved in the ionization processes.

Although photoelectron spectroscopy is often described as giving information on molecular structure, it is really concerned with the structures of positively charged ions. A simple example illustrates this; Fig. 3.18 gives a molecular orbital diagram for HCl. The molecule has four non-bonding π electrons, and two bonding σ electrons. There are therefore two ionization energies which can be measured: the π electrons have an ionization energy of about 13 eV, and the σ electrons one of about 16 eV. An alternative way of thinking about this is to say that the HCl^+ ion has two electronic states: a ground state $^2\Pi$, with electron configuration $\sigma^2\,\pi^3$, and an excited state $^2\Sigma$, with configuration $\sigma^1\,\pi^4$. The excited state lies about 3 eV above the ground state.

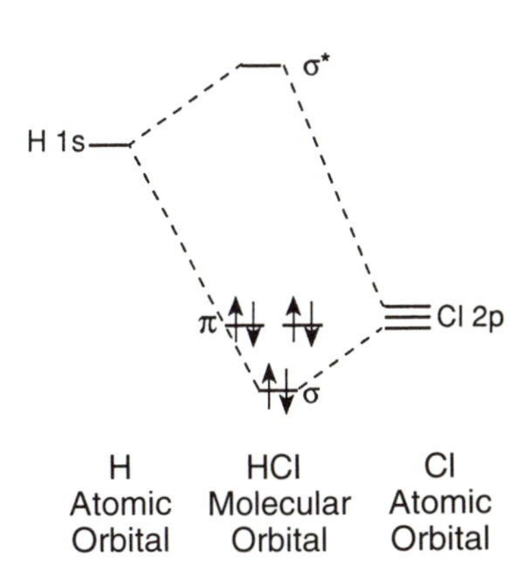

Fig. 3.18 Molecular orbital diagram for HCl.

At first sight these two descriptions might seem to be equivalent to each other; one considers the molecule, and the other the positive ion. This equivalence is only true however if the molecular orbitals remain broadly unaltered on ionization. This is not necessarily a very realistic model, as the size and shape of molecular orbitals depend on a balance between the electron–nuclear attraction and the electron–electron repulsion. If one electron is removed, the second term changes, and it need not affect all other orbitals equally. When an electron is removed, the other orbitals are said to relax, and the energy involved is called the relaxation energy.

In many cases the relaxation energy is not very important, and molecular orbital calculations, which predict orbital energies, can be of great value in assigning ionization energies. However the process is not infallible. The molecular orbital energies of the N_2 molecule have been shown to be as in Fig.3.19, with the highest occupied orbitals being 1π. We should therefore

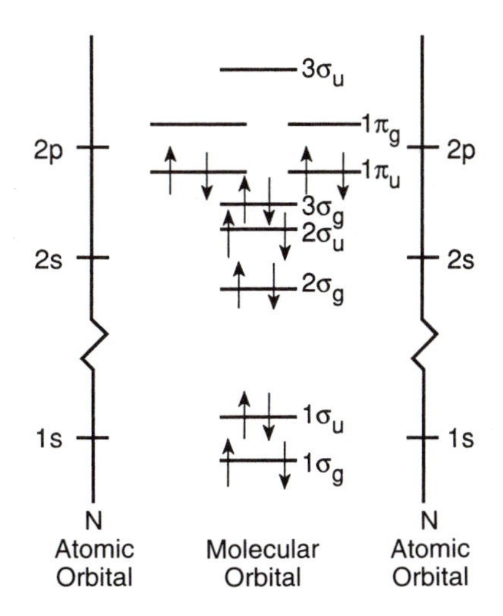

Fig. 3.19 The occupied molecular orbitals in the ground electronic state of N_2.

expect the N_2^+ ion to have a $^2\Pi$ ground state, with a $^2\Sigma$ excited state. In fact the reverse is the case, and the ground state is $^2\Sigma$. There is no dilemma here; we may write the configurations of N_2 and N_2^+ as:

$$N_2 \ldots\ldots \sigma^2 \pi^4$$

$$N_2^+ \ldots\ldots \pi^4 \sigma^1,$$

where the orbitals appear in order of energy. The π orbitals in the two cases are similar, but not identical, functions, and the ordering of the orbitals changes when the electron is removed. The important point is that photoelectron spectroscopy measures properties of the ion, not the molecule.

Photoelectron spectroscopy has also been applied to measure the ionization energies of core electrons; it is widely referred to as ESCA. Here an electron is removed from a 1s orbital; the ionization energies of these electrons are little altered when molecules are formed, and so ESCA can be used to identify individual chemical elements. Chemical shifts can be observed, however; the ionisation energy of a 1s electron depends to a small, but measurable, extent on the chemical environment of the atom.

3.13 Vibrational energy levels

In the last chapter we saw that diatomic molecules can vibrate, and that their vibrational motion is quantized. The allowed vibrational energy levels are roughly equally spaced, and the energy gap is such that at room temperature only the lowest level is significantly populated.

The situation is broadly similar in polyatomic molecules; although the vibrational motions of a polyatomic molecule may be more complicated, they can be considered as the sum of a number of simple vibrations. The actual number of such vibrations can be obtained from the number of atoms in the molecule. If the molecule has N atoms, then it will have $3N$ degrees of freedom, as each atom can move independently in three dimensions. Three of these degrees of freedom correspond to translation of the molecule along the three Cartesian axes, and three more correspond to rotation about these axes. The number of modes of vibration is therefore $(3N - 6)$. For a linear molecule there are only two axes about which it can rotate, and so the number of vibrational modes is $(3N - 5)$.

If we consider the CO_2 molecule, then it has four $(3 \times 3 - 5)$ vibrational modes. These are shown in Fig. 3.20, and are given their standard labelling,

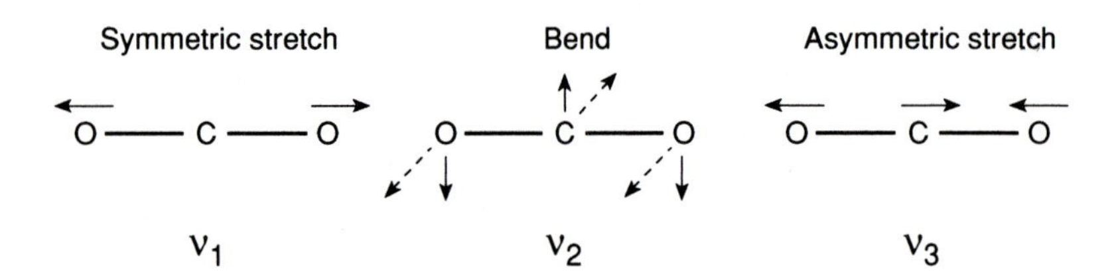

Fig. 3.20 The vibrational modes of CO_2.

ν_1, ν_2, and ν_3. Notice that ν_2 really corresponds to two separate, degenerate vibrations, one with the atoms moving into and out of the plane of the paper, and the other with the atoms moving just in the plane of the paper. They have the same energy and frequency. The modes can be labelled as involving bond stretching or bending.

For the H_2O molecule, there are now just three $(3 \times 3 - 6)$ vibrational modes; these are shown in Fig. 3.21. Note that in each case, the centre of gravity of the molecule is unchanged during the vibration.

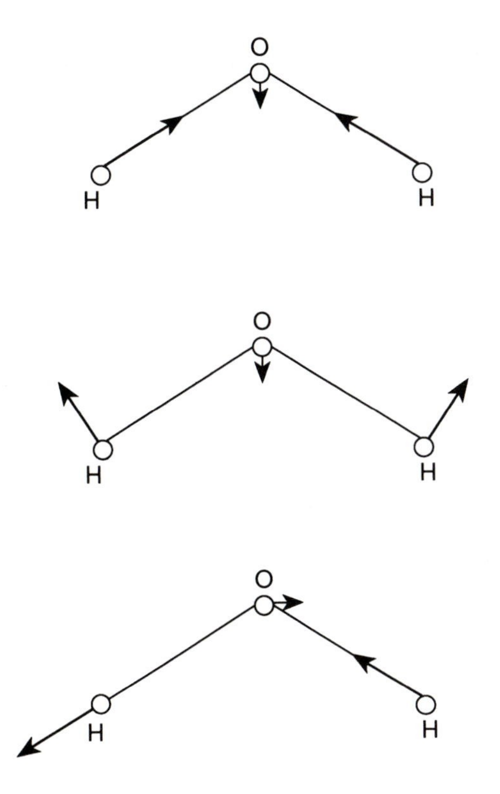

Fig. 3.21 The vibrational modes of H_2O.

To a good approximation, we can treat each vibration as a harmonic oscillator, just as in the diatomic case; the energy of vibration is then given by:

$$E = (v + \tfrac{1}{2})\, h\nu,$$

where ν is the vibration frequency. The energy required to promote the molecule from its vibrational ground state to its first excited state is then $h\nu$. Each vibrational mode will have its own characteristic frequency, depending on the force constant for the bond stretching or bending, and the masses of the atoms involved.

The assignment of frequencies to vibrational modes is too large a topic to be dealt with completely here, but there are two important points which can help in deciding which frequencies correlate with particular vibrations. Firstly, stretching frequencies tend to be higher than bending frequencies, as the force constants are higher for bond stretching; this is illustrated in the frequencies for CO_2, where the bending frequency ν_2 is 667 cm^{-1}, while the stretching frequencies are 1340 cm^{-1} (ν_1) and 2349 cm^{-1} (ν_3). Secondly, isotopic substitution can help in assigning vibrational frequencies. Thus the C–Cl stretching frequency is lower for ^{37}Cl than for ^{35}Cl, as the bonds have identical force constants but different atomic masses.

3.14 Characteristic frequencies

The number of vibrational modes increases quite rapidly with the number of atoms in a molecule, and it might therefore be thought that for large molecules the system was too complex for vibrations to be of great practical interest. In fact the reverse is true, and infra red spectroscopy, which detects vibrational transitions, is an established and versatile technique.

The chief value of infra red spectra arises from the fact that the vibrational frequencies of characteristic groups are relatively insensitive to the nature of the rest of the molecule. Thus the C=O group in $(CH_3)_2C{=}O$ has much the same vibrational frequency as that in $(C_2H_5)_2C{=}O$. Table 3.3 gives some characteristic frequencies; these are used to detect the presence of these groups in organic compounds. Note that, as we have seen before, the frequencies are expressed in wavenumbers, cm^{-1}. These values can be multiplied by c, the speed of light in cm s^{-1}, to give the frequency in Hz, and then by h, Planck's constant, to give an energy in J.

Table 3.3 Characteristic vibration frequencies

Group	Frequency (cm^{-1})
O–H (non-H-bonded)	3650–3600
N–H	3500–3200
=C–H	3100–3000
C–H	2970–2850
C=N	2275–2200
C=C–H	2260–2100
C=C	1680–1620
C–O	1780–1660
C–F	1400–1000
C–C	1250–700
C–Cl	800–600

The skeletal vibrations of organic molecules lie at rather lower energies in the region 700–1400 cm^{-1}. They often produce a complex vibrational spectrum, but even this complexity can be of value, acting as a fingerprint for the molecule.

3.15 Rotational energy levels

Diatomic molecules can rotate about their centre of gravity, and that rotation is quantized, with the energy levels depending on the molecule's moment of inertia; similarly polyatomic molecules can rotate, and their rotations are also quantized.

We can conveniently classify polyatomic molecules by their overall shape. A molecule has three principal moments of inertia, one about each perpendicular axis, I_A, I_B, and I_C. For a linear molecule $I_A = 0$, and $I_B = I_C$. For a spherical top, such as CCl_4, $I_A = I_B = I_C$. For a symmetric top such as CH_3Cl, $I_B = I_C \neq I_A$. Lastly asymmetric tops have $I_A \neq I_B \neq I_C$. The rotations of these types of molecule are discussed separately below.

Linear molecules

For linear polyatomic molecules, the energy levels are exactly as we encountered for diatomic molecules:

$$E = BJ(J+1),$$

where J is the rotational quantum number, and is an integer, and B is a constant whose value depends on the moment of inertia I:

$$B = h^2/8\pi^2 I.$$

The separation of two adjacent rotational levels is therefore:

$$\begin{aligned} E &= E_{J+1} - E_J \\ &= B(J+1)(J+2) - BJ(J+1) \\ &= 2B(J+1), \end{aligned}$$

so the rotational energy levels get further apart as J increases. If we can measure this separation, then we can calculate the value of the moment of inertia of the molecule; for the molecule OCS, $I = 1.38 \times 10^{-45}$ kg m^2.

In the case of a diatomic molecule, this measurement was enough to determine the bond length; in a polyatomic molecule, this is no longer so, as one piece of data cannot give values for all the bond lengths. The problem can be overcome by using isotopic substitution. If the value of I is measured for the molecules $OC^{32}S$ and $OC^{34}S$, then both bond lengths can be obtained. Although the masses of the atoms change, the bond lengths do not, as they depend only on the electronic structure of the molecule, and not the nuclear masses. The two molecules therefore have different values of I, and these two pieces of information allow two bond lengths to be calculated.

Spherical tops

For spherical top molecules, the rotation is again described by a quantum number J, which is integral; the energy of the allowed levels is given by:

$$E = BJ(J+1).$$

Once again the value of B depends on the molecule's moment of inertia: for the CCl_4 molecule, B is about 5 cm^{-1}.

Symmetric tops

In symmetric tops the molecule has two different moments of inertia, I_A and I_B: the third moment of inertia I_C is equal to I_B. If we define A= $h/8\pi^2 I_A$ and B= $h/8\pi^2 I_B$, then the energy of the symmetric top is given by:

$$E = BJ(J+1) + (A-B)K^2,$$

where J and K are both whole numbers. K represents the component of J along the molecular axis, and may take all values from $-J$ to $+J$. K thus describes the direction of the rotation; if $K = J$, then the rotation is about the molecular axis, while if $K = 0$, the rotation is end-over-end.

Asymmetric tops

Asymmetric tops are molecules with three different moments of inertia. They do have quantized energy levels, but these cannot be expressed in any simple form. In a number of cases their spectra have been analysed, however, and they give accurate information on bond lengths and angles, and on energy barriers to internal rotation.

Although the spacing of rotational energy levels clearly varies from molecule to molecule, the levels are typically much closer to each other than are vibrational or electronic energy levels. The spacings correspond to the energies of photons in the microwave region of the spectrum, and are typically comparable with kT at room temperature. It is therefore common form any rotational levels to be occupied, in contrast to vibrational and electronic levels, where only the ground state is normally occupied at room temperature.

3.16 Inversion in ammonia

The microwave spectrum of ammonia contains one further detail that we have not yet considered. It does not concern the rotational levels of the molecule,

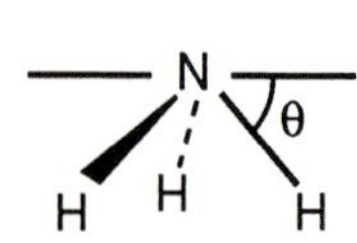

Fig. 3.22 Geometry of the NH_3 molecule.

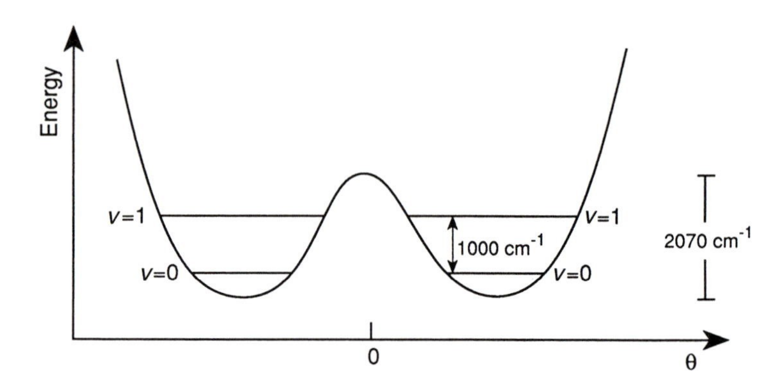

Fig. 3.23 Vibrational levels in NH_3 (without tunnelling).

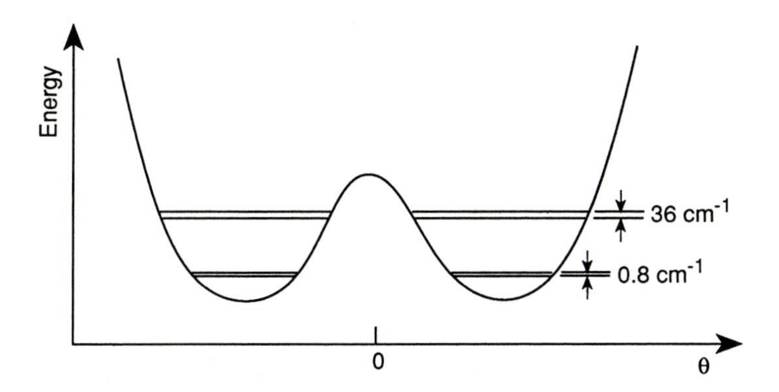

Fig. 3.24 Vibrational levels in NH_3 (including tunnelling).

although microwaves normally produce transitions between these levels; rather it is a feature of the vibrational levels.

The ammonia molecule is pyramidal, as shown in Fig. 3.22; its deviation from planarity is measured by the angle marked θ. If a graph is plotted of the potential energy of ammonia against θ, then a double minimum is obtained, as shown in Fig. 3.23. The height of the energy barrier is shown as 2070 cm^{-1}; classically this is the energy required to turn the molecule inside out, like an umbrella. Two vibrational energy levels are also shown.

If the vibrational wave functions are drawn in, it is found that because of the finite height of the barrier, the wave functions in the two halves of the diagram can overlap slightly. This is an example of quantum mechanical tunnelling. We must now consider linear combinations of the vibrational wave functions, just as we took linear combinations when we brought together two H atoms. These combinations will have slightly different energies, as shown in Fig. 3.24. The lowest vibrational state is therefore split into two levels, separated by 0.8 cm^{-1}; transitions between these levels give rise to the ammonia inversion spectrum.

4 Energy levels in NMR

NMR, or nuclear magnetic resonance, is perhaps the most important advance in molecular spectroscopy in the second half of the twentieth century. Together with the closely related ESR (electron spin resonance) spectroscopy the technique has had an impact in all areas of molecular science from chemistry to biochemistry and even medicine.

The energy levels involved in this form of spectroscopy are encountered even in simple atomic spectroscopy.

4.1 The effects of fields on atomic energy levels

An electric current going round a wire loop creates an effective magnet. Similarly an atom with some angular momentum has a rotating charge and so acts as a little magnet, but governed by the microscopic rules of quantum mechanics. The tiny magnet will line up with a magnetic field if, for example, placed between the poles of a magnet, just as a compass needle would. Different orientations of the magnet have different energies: the most stable lining up with the north pole pointing to the south pole of the external magnet and the least stable in the opposite direction—north to north and south to south. At the atomic level, not all orientations are possible, only those allowed by the rules of quantum mechanics.

If the total angular momentum of the atom is J, then different energies arise from orientations controlled by M_J, the component of J which has discrete values $J, J-1, \ldots, -J$ (i.e. $2J+1$ possible values of M_J).

In the case of the hydrogen atom ground state with the electronic configuration 1s, the atom is a $^2\mathrm{S}_{\frac{1}{2}}$ state, i.e. $J = \frac{1}{2}$, and M_J can be $+\frac{1}{2}$ or $-\frac{1}{2}$. These two levels have different energies in a magnetic field and their energy separation depends on the size of the applied magnetic field.

The nucleus of the hydrogen atom also has a spin $I = \frac{1}{2}$, and this too can act as an even smaller magnet with two orientations given by M_I values of $\pm\frac{1}{2}$.

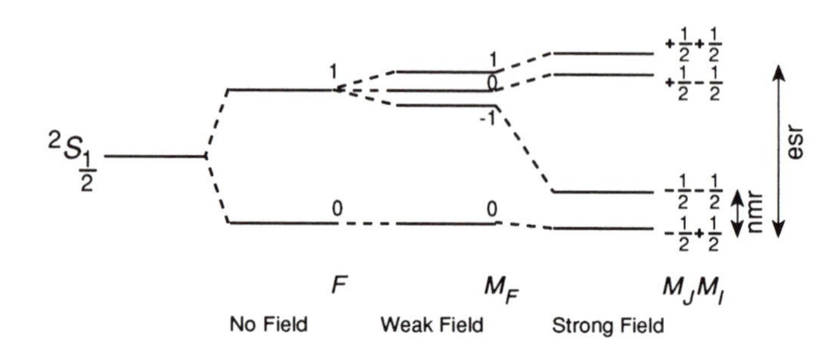

Fig. 4.1 NMR and ESR transitions in the H atom. In the strong field the nuclear and electron magnets are not coupled.

Thus, in a magnetic field the energy of the ground state of the hydrogen atom provides a rich pattern of energy levels as shown in Fig. 4.1. It is transitions between these levels which give rise to NMR and ESR spectra.

In molecules, for an NMR spectrum to be obtained it is necessary that the molecule should contain at least one nucleus which has a magnetic moment; as we shall see below, not all nuclei have magnetic moments, but most common elements have at least one isotope which can be used to obtain NMR signals. NMR has therefore been applied to a very wide variety of chemical systems.

Electron spin resonance is more restricted in its application, as all closed-shell molecules have only paired electrons and therefore no resultant electronic spin. ESR spectra can only be obtained from species with unpaired electrons; examples include organic and inorganic free radicals, and transition metal ions. They do not, however, need to be stable indefinitely at room temperature, and ESR has been used to study reactive intermediates.

The actual magnitudes of the energies of the transitions observed in magnetic resonance experiments depend on the strength of the applied magnetic field; with the field-strengths normally used in the laboratory, NMR signals are observed in the radio frequency region of the spectrum. ESR signals occur at higher energies, as the magnetic moments of electrons are greater than those of nuclei, and are observed in the microwave region.

If spin resonance experiments could detect only the presence of certain nuclei or of unpaired electrons in molecules, then they would be of some analytical interest, but no great importance. The real significance of spin resonance lies in the fact that nuclei and unpaired electrons also experience small magnetic fields from the motions of other nuclei and electrons, within their own molecules and sometimes in adjacent molecules. These fields modify the appearance of the spin resonance spectrum, which can then be used to obtain a very clear picture of the precise chemical environment in a molecule.

4.2 Basic principles of NMR

As we have seen, the necessary condition for a nucleus to give an NMR signal is that it should possess spin angular momentum, and hence have a magnetic moment. Each nucleus has a definite spin angular momentum whose value is determined experimentally. Table 4.1 shows the values for some common nuclei. As it usual in quantum mechanics, angular momentum is quantized in units of $h/2\pi$; the angular momentum is described by a quantum number I such that

$$\text{Angular momentum} = \sqrt{I(I+1)}\, h/2\pi,$$

where I is an integer or half-integer.

Although there is no simple theory which allows us to predict the value of I for every nucleus, some general rules can be formulated:

(a) Nuclei where the sum of the number of protons plus neutrons is odd have half-integral spin.

(b) Nuclei where the sum of the number of protons plus neutrons is even have integral spin.

Table 4.1 Nuclear spins of some common nuclei

Isotope	Per cent natural abundance	Spin I
^{1}H	99.984	$\frac{1}{2}$
^{2}H	0.016	1
^{3}H	0	$\frac{1}{2}$
^{13}C	1.108	$\frac{1}{2}$
^{14}N	99.635	1
^{17}O	0.037	$\frac{5}{2}$
^{19}F	100	$\frac{1}{2}$
^{31}P	100	$\frac{1}{2}$
^{33}S	0.74	$\frac{3}{2}$

(c) Nuclei with an even number of protons and an even number of neutrons have zero spin.

These rules are explained in part by the fact that both the proton and the neutron have spin angular momentum, with $I = \frac{1}{2}$. For example, the ^{13}C nucleus contains six protons and seven neutrons, and has $I = \frac{1}{2}$. The ^{2}H nucleus contains one proton and one neutron, and has $I = 1$; the ^{32}S nucleus contains sixteen protons and sixteen neutrons, and has $I = 0$. In particular we should note that both the ^{12}C and ^{16}O nuclei are non magnetic; this is of great importance in simplifying the ^{1}H-NMR spectra of organic compounds.

Nuclei with $I = 0$ have zero magnetic moments, but there is no simple way of predicting the magnetic moments of nuclei which do have spin angular momentum. The unit in which nuclear magnetic moments are usually measured is the nuclear magneton μ_N, which is equal to $eh/4\pi m_p$, where m_p is the mass of a proton; it has the value 5.05×10^{-27} J T^{-1} (1JT$^{-1} \equiv 1$ m^2 A). The magnetic moment μ is then given by

$$\mu = g\, \mu_N \sqrt{I(I+1)},$$

where g is a numerical factor whose value has to be determined experimentally for each nucleus. g values are usually, though not invariably, positive. The relationship between the angular momentum p and the magnetic moment μ is therefore

$$\mu = (2\pi/h)\, g\, \mu_N p.$$

The ratio of the magnetic moment to the spin angular momentum is called the magnetogyric ratio, γ:

$$\gamma = 2\pi \frac{g\, \mu_N}{h}.$$

When a nucleus is placed in a magnetic field, it experiences a torque which causes is to precess about the field direction, just as a gyroscope precesses in a gravitational field (Fig. 4.2). The frequency of this precession depends on the applied field and is called the Larmor frequency. The angle at which the nucleus precesses may not take any value, but, as is usual in quantum mechanics, is such that the component of the angular momentum in the direction of the field is $(h/2\pi)M_I$, where M_I is another quantum number and may take the values $I, (I-1), \ldots, (1-I), -I$. Thus for a proton, $I = \frac{1}{2}$, so $M_I = +\frac{1}{2}$ or $-\frac{1}{2}$, which we may speak of roughly as the nucleus being aligned with or against the field. For a deuterium nucleus, $I = 1$, and so three orientations of the nucleus are possible, with M_I values of $+1$, 0, and -1.

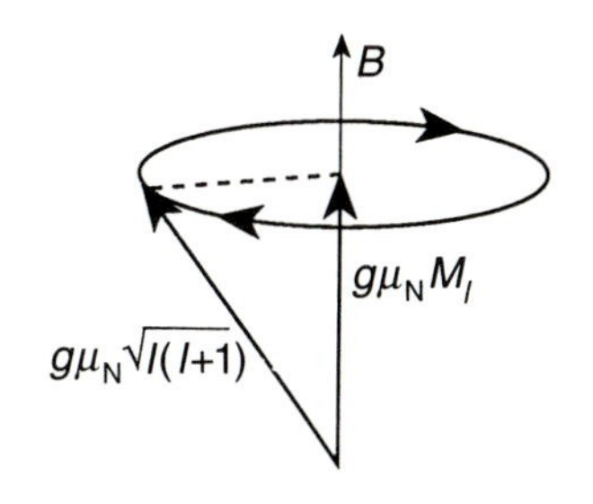

Fig. 4.2 Precession of the nuclear magnet in a magnetic field.

The potential energy of a nuclear magnetic moment μ in an external field of flux density B is given by

$$E = -B\mu\cos\theta,$$

where θ is the precession angle; if B is measured in teslas then E is in joules. The quantity $\mu\cos\theta$ is simply the component of the magnetic moment in the field direction, which is the component of the angular momentum in the field direction multiplied by the magnetogyric ratio. Therefore,

$$\begin{aligned} E &= -B\,h\,M_I\,\gamma/2\pi \\ &= -B\,M_I\,g\,\mu_N. \end{aligned}$$

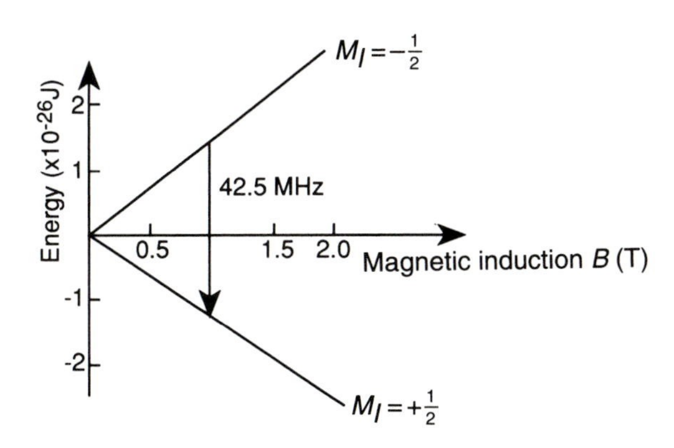

Fig. 4.3 Energies of the two states of the proton in a magnetic field.

Figure 4.3 shows the energies of the two states of the ^{1}H nucleus in a magnetic field. Transitions between these two states are magnetic-dipole-allowed, and the transition energy is

$$\begin{aligned} \Delta E &= B\,g\,\mu_N\left[\tfrac{1}{2} - \left(-\tfrac{1}{2}\right)\right] \\ &= B\,g\,\mu_N. \end{aligned}$$

This is the energy required to reverse the direction of the nuclear spin. The frequency of the radiation corresponding to this energy is

$$\nu = \frac{B\, g\, \mu_N}{h},$$

which is equal to the Larmor precession frequency. For a proton in a field of 1 T, the frequency is

$$\begin{aligned}\nu &= \frac{1 \times 5.585 \times (5.05 \times 10^{-27})}{(6.63 \times 10^{-34})}\ \text{Hz} \\ &= 4.25 \times 10^{7}\ \text{Hz} \\ &= 42.5\ \text{MHz},\end{aligned}$$

which is in the radiofrequency region of the spectrum.

Similar calculations can be carried out for other nuclei; when $I > \frac{1}{2}$, the selection rule for transitions is $\Delta M_I = +1$. As each nucleus has its own g value, so it has its own resonant frequency at any given field strength. Typical values for some common nuclei are given in Table 4.2; as values of g vary quite widely, so do the resonant frequencies, and any NMR experiment detects the signals from one type of nucleus only.

Table 4.2 NMR frequencies at 1 T

Nucleus	NMR frequency (MHz)
^{1}H	42.576
^{2}H	6.536
^{3}H	32.434
^{13}C	10.705
^{14}N	3.076
^{17}O	5.772
^{19}F	40.054
^{31}P	17.238
^{33}S	3.266

One of the practical problems of NMR is that it is inherently an insensitive technique. When a sample is irradiated at the resonant frequency, transitions are induced in which nuclei parallel to the field become anti-parallel, and vice versa. These two processes, stimulated absorption and stimulated emission, are of equal probability, and a net absorption of energy is detected only because one orientation has a greater equilibrium population than the other. The energy differences in NMR are so small (compared with kT) that the numbers of nuclei in each orientation are almost exactly equal. For example, for a ^{1}H nucleus with a resonant frequency of 60 MHz, the Boltzmann distribution law gives

$$
\begin{aligned}
n_+/n_- &= \exp \frac{-h\nu}{kT} \\
&= \exp - \frac{(6.63 \times 10^{-34}) \times (60 \times 10^6)}{(1.38 \times 10^{-23}) \times (300)} \\
&= \exp - (1 \times 10^{-5}) \\
&= 0.99999.
\end{aligned}
$$

The observation of an NMR signal depends on this slight excess of 1 part in 10^5 of one spin state over the other; but for most purposes in the interpretation of spectra, it is clearly safe to assume that all nuclear spin states are equally populated.

The intrinsic probability of an NMR transition is independent of the chemical environment of a nucleus, and the intensity of an NMR line is, under optimum conditions, proportional to the number of nuclei which give rise to the transition. NMR may thus be used in quantitative analysis. For example, by comparing the signal intensity with that of a standard, the water content of silica balls, used as catalysts, may be determined from their proton absorption. Alternatively, the relative intensities of lines in a spectrum can be measured; in this way the aliphatic–aromatic ratios in oil mixtures can be determined, because aliphatic and aromatic hydrogen nuclei resonate, as we shall see below, at slightly different frequencies.

In NMR the sample is usually in the liquid state, either pure or in solution. It is also possible to use solid samples, although as we shall see later, the appearance of the spectra of solids is markedly different from that of liquids.

The first nucleus to be used widely in NMR experiments was ^1H, the proton; this isotope has a high natural abundance, and hydrogen is found in a wide variety of compounds. Later other nuclei such as ^{19}F, ^{23}Na, and ^{31}P were employed, and as experimental techniques improved it became possible to use isotopes such as ^{13}C, where the natural abundance is only 1 per cent.

4.3 Chemical shifts

In practice it is found that all the nuclei of any given isotope do not resonate at exactly the same frequency; the resonant frequency is affected slightly by the chemical environment of the nucleus. Fig. 4.4 shows the proton NMR spectrum of $HCOOCH_3$; the spectrum is plotted with the frequency held constant and the magnetic field strength being varied to bring all the nuclei to resonance in turn. The spectrum consists of two peaks, corresponding to the $-CH_3$ and $-CHO$ protons; the intensities of the two peaks are in the ratio of 3:1, reflecting the relative numbers of each sort of proton.

The splitting of the NMR spectrum into two peaks is caused by the fact that each nucleus is partially shielded from the external field by the electrons which surround it. The external field causes these electrons to move and their motion sets up a new field, which by Lenz's law opposes the original field. The magnitude of this shielding depends on the local electron density, and so nuclei in different environments are shielded to different extents, that is have different 'chemical shifts'.

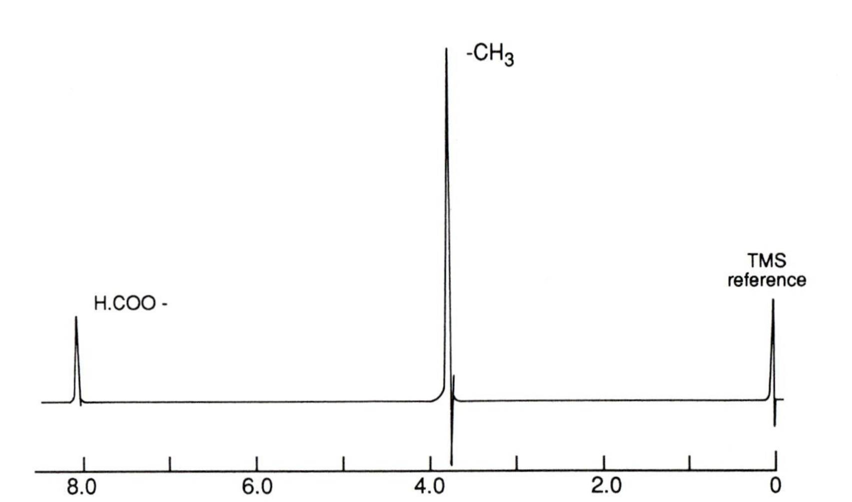

Fig. 4.4 NMR spectrum of $HCOOCH_3$ with the delta scale in ppm.

As chemical shifts are very small in comparison with the fields used in spectrometers, it is normal to measure the position of a peak relative to that of an arbitrarily chosen standard. Tetramethylsilane (TMS), $(CH_3)_4Si$, is added to samples to act as a standard for both ^1H and ^{13}C spectra, as it is very unreactive and gives a single peak which does not normally overlap with other peaks.

Chemical shifts are proportional to the field at which a spectrometer operates. There is no standard field-strength at which all spectrometers operate and so chemical shifts are measured not in field or frequency units, which would be machine-dependent, but as a dimensionless ratio. The chemical shift of a nucleus i, δ_i, is given by

$$\delta_i = \frac{B_{TMS} - B_i}{B_{TMS}},$$

where B_i and B_{TMS} are the fields required to bring i and TMS to resonance. To bring the numbers to a convenient range, the δ values are then multiplied by 10^6, and expressed as parts per million (ppm). The range of chemical shifts found for ^1H nuclei covers about 15 ppm; the ^{13}C nucleus has a higher electron density and its range of chemical shifts is about 300 ppm.

Chemical shifts for H nuclei are also sometimes expressed on the τ scale; here TMS is assigned a shift of 10 ppm and peaks at lower fields have lower τ values. The τ and δ scales are related:

$$\delta = 10 - \tau,$$

as shown in Fig. 4.5, the spectrum of CH_3COOH. Notice that the proton attached to the electronegative oxygen atom has a relatively low electron density, and is therefore not shielded as much as the $-CH_3$ protons, with a higher electron density.

Chemical shifts are often the most useful features of NMR spectra. A very wide variety of organic compounds have been studied by NMR spectroscopy;

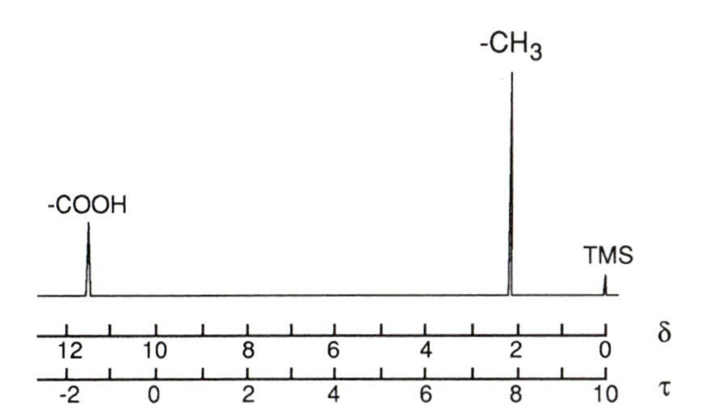

Fig. 4.5 ^{1}H spectrum of CH_3COOH (in $CDCl_3$).

at first ^{1}H-NMR was used exclusively, but more recently ^{13}C spectra have become more routinely available. It is found that many functional groups have a characteristic chemical shift, whose value is affected relatively little by substituents; some examples are given in Table 4.3. The use of such tables has made NMR a particularly powerful technique in the analysis of unknown compounds; it can often reveal structural features which are difficult to detect by other methods.

Table 4.3 Chemical shifts for CH_3X

X	τ
$-Si(CH_3)_3$	10
–aliphatic group R	9.1
$-CR{=}CR_2$	8.3
–COR	7.9
–I	7.84
$-NR_2$	7.8
–aromatic hydrocarbon	7.7
–Cl	6.95
$-N^{+}R_3$	6.7
–OCOR	6.3
–F	5.74
$-NO_2$	5.72

The chemical shifts of many functional groups may be rationalized by simple electronegativity arguments; thus the introduction of a fluorine atom into a group decreases the electron density, producing less shielding and a shift to lower field. Deviations from these simple predictions are often found in aromatic systems, where the external magnetic field causes the mobile π electrons to circulate. These 'aromatic ring currents' are shown in Fig. 4.6; they cause protons attached directly to the benzene ring to resonate at low fields, as the field from the ring reinforces the external field outside the ring. However, protons immediately above or below the ring would resonate at high fields, as here

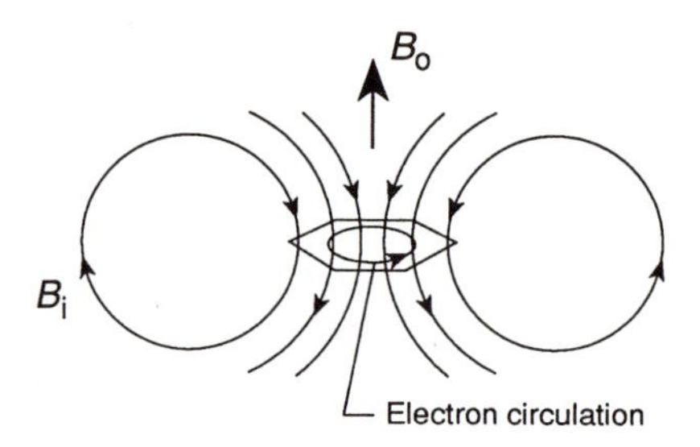

Fig. 4.6 Aromatic ring currents.

the field from the ring currents opposes the external field. This effect has been observed in the cyclophanes,

$$\begin{array}{ccc} CH_2\text{–}C_6H_4\text{–}CH_2 \\ | \qquad\qquad | \\ CH_2\text{–}(CH_2)_n\text{–}CH_2 \end{array}$$

where the CH_2 protons show a wide variety of chemical shifts, depending on their position relative to the benzene ring. Ring current shifts are often large and have been used to obtain information on the conformations of biochemical molecules; the resonances of some groups are found at unusually high fields, which can be interpreted in terms of their proximity to an aromatic ring in some other part of the molecule.

Chemical shifts are also strongly affected by hydrogen bonding. Hydrogen bonding shifts the proton resonance to low field; the effect can be large, of the order of 10 ppm. The hydrogen bonding may be either intramolecular, as in nitrophenol, or intermolecular. Chemical shifts in hydrogen-bonded systems are often strongly concentration-dependent; this dependence can give useful information on the hydrogen-bonding, but means that lines from –OH and –NH– protons have to be interpreted with great care in structure determination.

4.4 Spin–spin coupling

Many NMR spectra contain more lines than can be explained on grounds of chemical shift differences alone; this further structure arises from the interactions between the magnetic nuclei within a molecule. Figure 4.7 shows the ^{1}H-NMR spectrum of $CHCl_2-CH_2Cl$; absorption occurs in two different parts of the spectrum, corresponding to the two hydrogen environments, but each of these absorptions is split into a series of well-resolved lines.

There is a direct interaction between any pair of magnetic dipoles in a molecule, but in liquid samples the rapid molecular tumbling which occurs averages this interaction to zero. However, two magnetic nuclei can also interact via the bonding electrons between them; this interaction is not averaged to zero by molecular tumbling and gives rise to the fine structure visible in Fig. 4.7. We can illustrate this mechanism by considering a ^{13}C and a ^{1}H nucleus joined by a

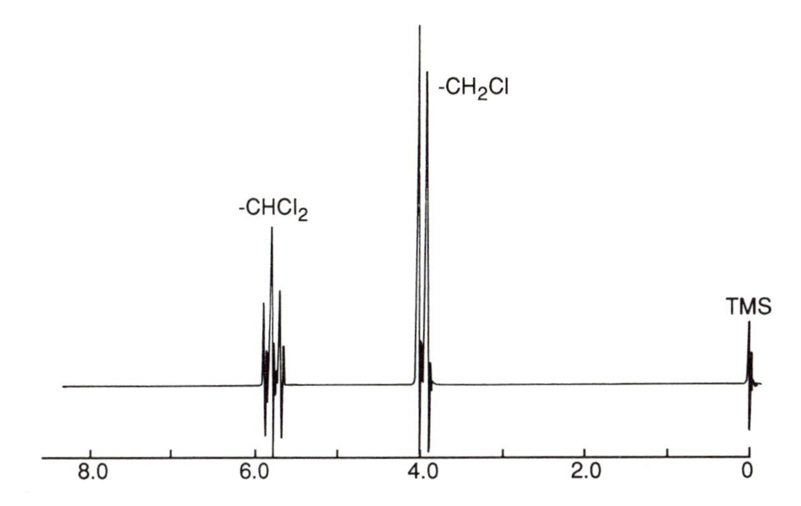

Fig. 4.7 ^{1}H spectrum of $CHCl_2-CH_2Cl$ on the delta scale.

single covalent bond (Fig. 4.8). The bond consists of two electrons whose spins (and therefore magnetic moments) are opposed—by the Pauli exclusion principle. Each nucleus attracts both electrons electrostatically, but there is a further magnetic interaction, which is favourable if the electron and nuclear magnetic moments are parallel to each other. If the ^{13}C and ^{1}H nuclei have opposed spins, then one electron will move fractionally closer to the ^{13}C nucleus and the other to the ^{1}H nucleus, so that both magnetic interactions are favourable. This cannot occur if the ^{13}C and ^{1}H nuclei are parallel; if one electron has a favourable interaction with one nucleus, then the other electron, at the opposite end of the bond, must have an unfavourable interaction with the other nucleus. The parallel arrangement of nuclear spins therefore has a higher energy than the paired arrangement.

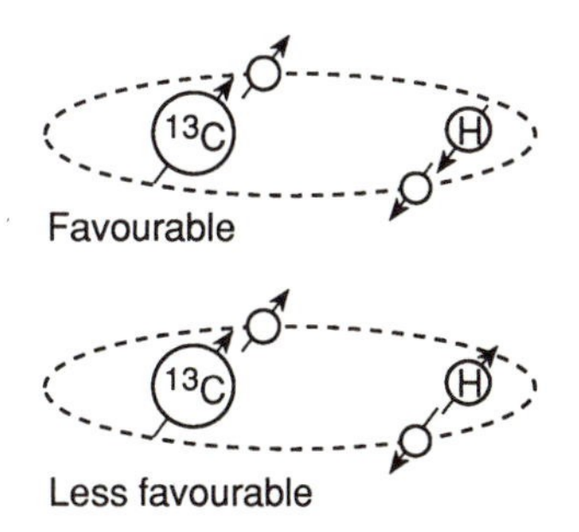

Fig. 4.8 The coupling of nuclear spins via bonding electrons.

The effect of this coupling on the ^{1}H spectrum is to produce two lines of equal intensity, corresponding to the two possible orientations of the ^{13}C nucleus. The magnitude of the splitting is not dependent on the magnetic field strength and is usually measured in frequency units (Hertz) and not parts per million. The splitting is called the spin–spin coupling constant, J. The coupling has an identical effect on the ^{13}C spectrum; two lines of equal intensity are seen, corresponding to the two orientations of the ^{1}H nucleus, and the splitting is the same as in the ^{1}H spectrum.

In fact, of course, the splitting of the energy levels does not necessarily split the peaks, but a more careful analysis does show how this arises. The spin energy of two nuclei, A and X, of spin quantum number $M_I(\mathrm{A})$ and $M_I(\mathrm{X})$ respectively is given by

$$\frac{E_{\mathrm{spin}}}{h} = -(M_I(\mathrm{A})\nu_A + M_I(\mathrm{X})\nu_{\mathrm{X}}) + M_I(\mathrm{A})M_I(\mathrm{X})J_{\mathrm{AX}},$$

where J_{AX} is the spin–spin coupling constant between the nuclei. There are thus four possible values for the spin energy, depending on the various values

Table 4.4 Spin energies for combinations of nuclear spin

$M_I(A)$	$M_I(X)$	Label	E_{spin}/h
$+\frac{1}{2}$	$+\frac{1}{2}$	$\alpha\alpha$	$-\frac{1}{2}(\nu_A+\nu_X)+J_{AX}/4$
$+\frac{1}{2}$	$-\frac{1}{2}$	$\alpha\beta$	$-\frac{1}{2}(\nu_A-\nu_X)-J_{AX}/4$
$-\frac{1}{2}$	$+\frac{1}{2}$	$\beta\alpha$	$+\frac{1}{2}(\nu_A-\nu_X)-J_{AX}/4$
$-\frac{1}{2}$	$-\frac{1}{2}$	$\beta\beta$	$+\frac{1}{2}(\nu_A+\nu_X)+J_{AX}/4$

of the spin quantum numbers. These four values are usually labelled as $\alpha\alpha$, $\alpha\beta$, $\beta\alpha$, and $\beta\beta$, as in Table 4.4.

There are two transitions corresponding to absorption of energy by nucleus A, for which $\Delta M_I(\text{A}) = -1$, $\Delta M_I(\text{X}) = 0$. These are $\beta\alpha \rightarrow \alpha\alpha$, for which $\Delta E/h = \nu_\text{A} - \frac{1}{2}J_\text{AX}$ and $\beta\beta \rightarrow \alpha\beta$, for which $\Delta E/h = \nu_\text{A} + \frac{1}{2}J_\text{AX}$. If X did not exist, the peak would be at ν_A. Spin–spin coupling causes this peak to split into two, one peak $\frac{1}{2}J_\text{AX}$ above, and the other $\frac{1}{2}J_\text{AX}$ below, the original.

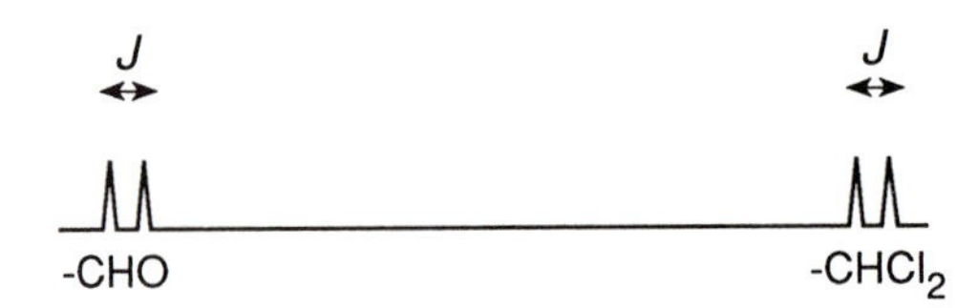

Fig. 4.9 Spin–spin coupling in the proton NMR spectrum of $CHCl_2CHO$.

It is not necessary that the two nuclei should be directly bonded to each other; spin–spin coupling is seen between the ^{1}H nuclei in $CHCl_2$–CHO, as in Fig. 4.9. Note that the splittings of the two pairs of lines are equal. Generally it is found that spin-spin coupling constants decrease as the internuclear distance increases; some typical values are given in Table 4.5. The sign convention is that J is positive where the anti-parallel arrangement of nuclei is more stable and negative where it is less stable.

Table 4.5 Spin–spin coupling constants (Hz)

^{1}H–^{1}H	280
^{1}HC–C^{1}H	6–8
^{1}HC–C–C^{1}H	1–2
^{1}HC–C–C–C^{1}H	< 0.5

Many molecules contain more than two magnetic nuclei which can spin-couple with each other; we can distinguish a number of situations, each with its own characteristic splitting pattern.

A particularly simple spectrum arises where two or more nuclei are in exactly equivalent chemical environments, such as the three protons in a $-CH_3$ group. Although the magnetic moments of the nuclei do interact, it can be shown that all transitions allowed by the selection rules occur at exactly the same energy, and so only one line is observed in the spectrum. This simplifies the appearance of many spectra and is summarized in the rule that spin-splitting is not observed between equivalent nuclei.

A more complex splitting pattern is produced when one proton couples to two other equivalent protons, as in CH_2Cl–CHO. If we consider the magnetic environment of the –CHO proton, there are now three possibilities. Both the protons in the $–CH_2Cl$ group could be parallel to the field, both could be anti-parallel, or one could be parallel and the other anti-parallel. The third possibility is twice as likely as the other two, as there are two ways in which it can be achieved. The absorption of the –CHO proton is therefore split into three lines, with intensities in the ratio of 1:2:1. The $–CH_2Cl$ protons couple to the single –CHO proton, and so give rise to just two lines, with a different chemical shift.

If a nucleus interacts with three equivalent spins, then four lines are obtained in the spectrum, with intensities in the ratio of 1:3:3:1.

In all the analyses above, we have treated spin-splitting as a small perturbation superimposed on the general pattern of chemical shifts in a spectrum. This is accurate, provided that the chemical shift between the nuclei which couple is much greater than the spin-coupling constant. If the spin-coupling constant and the chemical shift are of comparable magnitude, then deviations occur, both in the positions of the lines and their intensities; this is illustrated in Fig. 4.10.

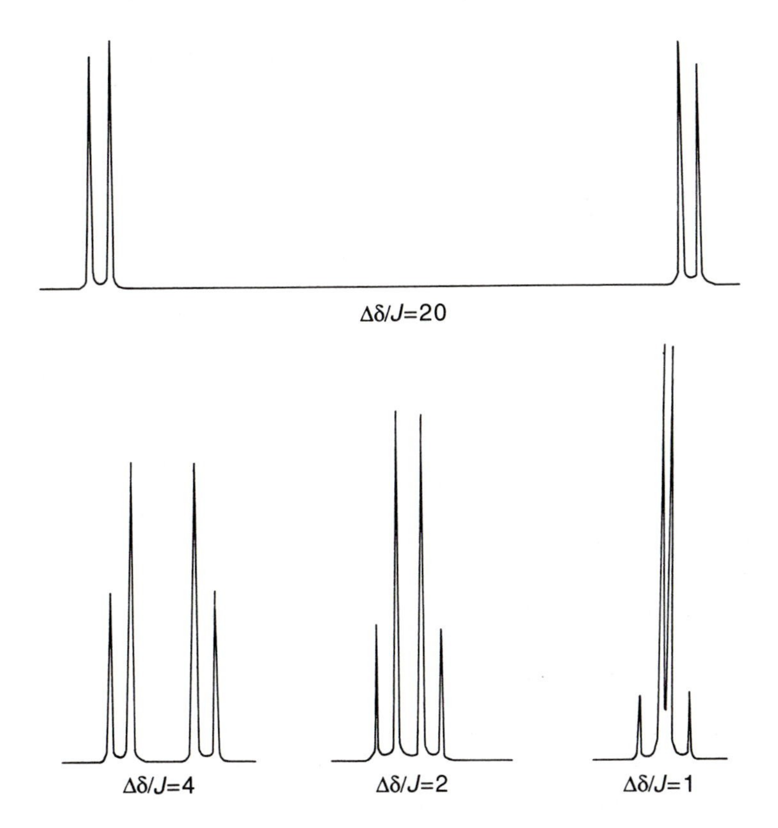

Fig. 4.10 Computer-simulated spectra for the interaction of two protons.

As the chemical shift increases with applied field, whereas the spin-coupling constant does not, these deviations are particularly marked when spectrometers operate at low magnetic fields. This is one reason why the present trend is to build spectrometers operating at very high fields.

The relative abundances of the ^1H and ^{13}C nuclei have an important effect on the spin coupling which is observed in the NMR spectrum of organic compounds. Virtually all naturally occurring hydrogen is ^1H, whereas ^{13}C represents only about 1 per cent of all naturally occurring carbon, the rest being the non-magnetic ^{12}C. In consequence, all ^1H-NMR spectra show spin coupling between pairs of hydrogen atoms, but not between hydrogen and carbon atoms, as almost all ^1H nuclei are bonded to non-magnetic carbon atoms. Similarly, ^{13}C spectra show spin coupling between carbon and hydrogen atoms, as all carbon atoms are bonded to magnetic hydrogen atoms, but not between pairs of carbon atoms, as almost all ^{13}C nuclei are bonded to non-magnetic ^{12}C nuclei. Isotopic enrichment can, of course, be used to obtain carbon–carbon coupling constants.

The appearance of a spin multiplet can be used to give important information on the structure of an unknown molecule. The most obvious parameter is the number of lines in the multiplet and the intensity distribution, giving direct information on the number of interacting nuclei. Thus the appearance of a 1:3:3:1 quartet in a ^{13}C spectrum could indicate the presence of a $-CH_3$ group. In practice, this is not always as useful as it might be thought. Lines of low intensity can be difficult to distinguish from background noise, and intensity distributions do not necessarily follow the simple rules, so the appearance of a spin multiplet is not always conclusive evidence on its own.

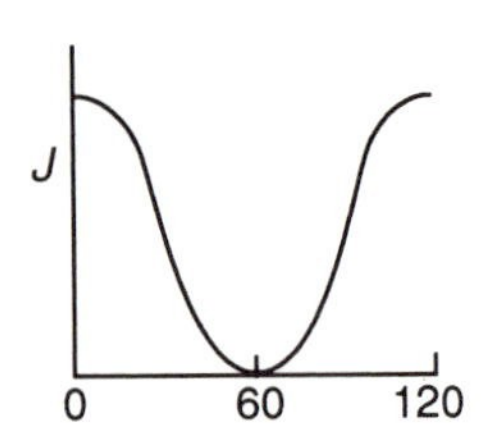

Fig. 4.11 Variation of spin–spin coupling with dihedral angle.

The magnitudes of spin–spin coupling constants can be of value in structure determination. These constants have values which are often characteristic of the stereochemistry of a molecule: for example the J value for two protons in a $-CH{=}CH-$ group is about +10 Hz if the protons are *cis*, and about +18 Hz for the *trans* arrangement. Measurement of spin splitting therefore allows the stereochemistries of substituted alkenes to be determined. Coupling constants can also be used to give conformational information in more flexible systems; for example, the value of J for two protons in a $>CH-CH<$ fragment depends strongly on the dihedral angle between the two C–H bonds (see Fig. 4.11). Although the precise shape of the graph depends on the nature of the substituents, the method is accurate enough to distinguish conformations in saturated ring compounds.

4.5 Rate process

One of the most remarkable features of NMR is the extreme narrowness of the lines; it is quite common to observe lines of width less than 1 Hz when the resonant frequency is of the order of 10^8 Hz. The uncertainty principle requires that, for a line to have a width less than 1 Hz, the energy state of the nucleus must remain constant for a time greater than of the order of 1 s. Many chemical processes occur faster than this, and they can modify the appearance of an NMR spectrum.

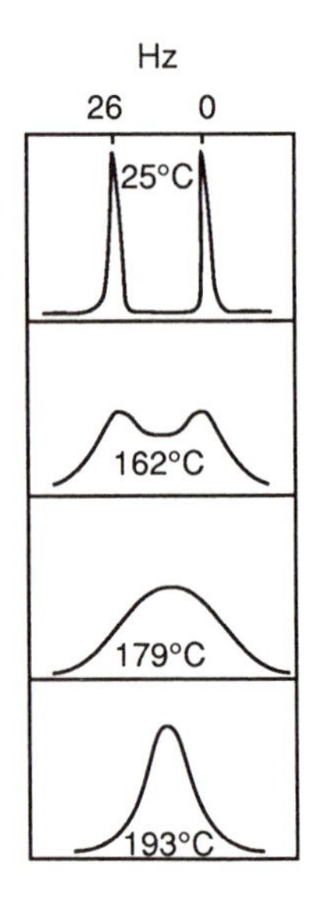

Fig. 4.12 Effect of temperature on the ^1H spectrum of $(CH_3)_2N-N{=}O$.

We can illustrate this effect by considering the molecule $(CH_3)_2N-N{=}O$, whose proton spectrum is shown in Fig. 4.12. At room temperature rotation about the N–N bond is very slow and the two $-CH_3$ groups are in different chemical environments, which one *cis* and the other *trans* to the $-N{=}O$ bond. The spectrum consists of two sharp lines. The rate of rotation about the N–N bond increases with temperature, and as the average lifetime of the molecule in one energy state decreases, so the linewidth increases. As the temperature continues to rise, the line broadening becomes comparable with the chemical shift difference between the two $-CH_3$ groups and the peaks begin to merge. However, as the temperature rises further, a single peak is formed, whose width decreases with further temperature increases. Here the rate of rotation has become rapid compared with the chemical shift difference (expressed in Hertz), and the chemical shift of this line is an average of the chemical shifts of the conformations of the $-CH_3$ groups.

Narrow lines are obtained in NMR spectra when a nucleus undergoes changes in its environment which are very slow—in which case each conformation gives rise to a line—or when it undergoes very rapid changes, when only a single line, representing its average environment, is seen. The key question here is what is meant by fast and slow. Processes with half-lives greater than 1 s do not normally give measurable broadening of lines and are therefore 'slow'. Similarly, chemical shift differences are normally less than 10^3 Hz and so processes with half-lives much less than 10^{-3} s are 'fast'. Processes with intermediate lifetimes, of the order of 10 ms, have a marked effect on NMR lineshapes and widths, and analyses of spectra have given detailed information on the kinetics of a variety of chemical reactions and restricted rotations.

Common examples of very rapid chemical exchanges include the interconversion of chair forms in cyclohexane rings and proton exchange in acidic solutions. In cyclohexane it is known that each hydrogen atom occupies either an equatorial or an axial position in the ring, but the rate at which one chair form flips into the other form—thereby exchanging axial and equatorial forms—is such that only one line is seen in the NMR spectrum at room temperature, at a position halfway between those expected for an equatorial and axial proton respectively. Similarly, the proton spectrum of CH_3COOH in water shows only two lines, one from the $-CH_3$ group and one from the $-COOH$ and H_2O protons. These exchange rapidly: $CH_3COOH + H_2O \rightleftharpoons CH_3COO^- + H_3O^+$, and it is not possible to observe separate lines for the $-COOH$ and H_2O environments. When a nucleus moves very rapidly between two non-equivalent positions, the value of the 'average' chemical shift can sometimes be used to obtain the equilibrium constant, if the chemical shifts of the two positions separately are known.

A further consequence of rapid chemical exchange is the collapse of spin–spin coupling. In the proton spectrum of pure CH_3OH, the $-CH_3$ absorption is split into two lines by the neighbouring $-OH$ group. When mineral acid is added, the $-OH$ proton undergoes rapid exchange, which causes fast reorientation of its magnetic moment. The $-CH_3$ group now experiences only the 'average' orientation of the $-OH$ proton, and its signal collapses to a single line.

4.6 Double resonance

The effects of spin–spin coupling in a spectrum may also be removed by irradiating the sample with a strong radiofrequency field. This is one example of double resonance, in which two magnetic nuclei are simultaneously irradiated at their resonant frequencies.

In spin decoupling the sample is irradiated strongly with the resonant frequency of one of the magnetic nuclei. This causes this nucleus to undergo very rapid reorientation in the magnetic field, with the result that adjacent nuclei no longer experience spin coupling and their resonances collapse to a single line. This can be very desirable when a nucleus is coupled to several other nuclei, giving a complex pattern of lines which may overlap strongly with each other; analysis may only be possible when spin decoupling is used. The technique can also be used to establish which nuclei in a molecule couple together, and hence to obtain structural information. Spin decoupling can also give information on the relative signs of spin-coupling constants; only their magnitudes can be deduced from simple splitting patterns.

A particularly important example of spin decoupling is the use of broadband decoupling in ^{13}C-NMR. Here the sample is irradiated strongly over a range of frequencies covering all proton absorptions, and the ^{13}C spectrum is then observed normally. The effect is to remove all the proton fine structure from the spectrum; this structure is of little analytical value and the intensity of each nucleus is concentrated into a single line. The low natural abundance of ^{13}C means that no carbon-carbon spin coupling is observed, and so ^{13}C spectra are especially free of complexity.

Another type of double-resonance experiment is 'spin tickling'. It is similar to spin decoupling, but the irradiation used is much weaker. The weak irradiation of a single resonance line splits all transitions which share a common energy level with the line into doublets. Spin tickling can be of great value in the assignment of complex spectra; it can also be of help in determining the signs of J values.

4.7 NMR of solids

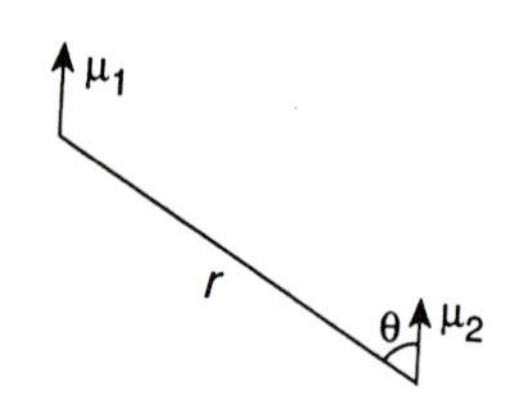

Fig. 4.13 The interaction of magnetic dipoles.

All the examples that we have considered so far have involved liquid samples, where there is rapid molecular tumbling. This tumbling removes from the spectra the effects of the direct dipolar interaction between magnetic nuclei, which is averaged by the motion to zero. However, in solid samples there is no molecular tumbling, and the direct interaction of magnetic dipoles now dominates the spectra, which consist of broad bands rather than a series of sharp lines.

The interaction energy of two magnetic dipoles separated by a distance r and at an angle θ to the line connecting them (Fig. 4.13) is given by

$$E = \frac{k(3\cos^2\theta - 1)}{r^3}.$$

The value of k depends on the size of the dipoles and whether the nuclei are equivalent or not, but for two protons in the same molecule, the value of E may be of the order of 50 kHz, when expressed in frequency units. This interaction energy is far greater than those involved in chemical shifts and spin–spin coupling.

The NMR spectrum of a molecular solid containing two magnetic nuclei consists of two peaks, as the energy of one nucleus depends on the orientation of the other. The separation of the peaks is large and the peaks themselves are broad, because of unresolved interactions with more distant nuclei. The peak separation is a function of the angle θ, and so for a single crystal the NMR spectrum varies if the crystal is rotated in the magnetic field. In a polycrystalline material, all possible values of θ are represented, and the resulting spectrum is an envelope. The width of the spectrum now depends only on the value of $1/r^3$; measurement of spectra is an important method of determining the hydrogen–hydrogen distances in solids, which are not easily measured by diffraction methods.

Similar results are obtained from crystals with more than two magnetic nuclei; the precise shape of the bands depends on the number and relative positions of the nuclei. Observation of band shapes can therefore be used to identify what species are present in a solid: for example, NMR shows that solid hydrated HNO_3 contains H_3O^+ ions, with an equilateral triangular arrangement of hydrogen atoms, rather than separate H_2O and HNO_3 molecules.

Studies of linewidths of solid NMR spectra can also give information on molecular motion in solids. If NMR spectra are measured as a function of temperature, it is often found that the linewidth undergoes marked changes at temperatures well below the melting point. These changes can be attributed to the onset of some motion within the crystal lattice; for example, the spectrum of benzene narrows appreciably at around $-170°C$, and this is interpreted as showing that above this temperature the molecules are free to rotate within the crystal lattice.

Dipolar broadening so dominates the NMR of solids that no details of chemical shifts can be seen. These details are precisely what makes liquid state NMR such a sensitive tool for determining molecular structure, and recently considerable efforts have gone towards measuring solid-state spectra with the dipolar broadening removed. This can be achieved by 'magic-angle spinning', in which the sample is spun very rapidly at an angle of $54° \, 44'$ to the magnetic field (Fig. 4.14). As we saw above, the interaction between the two magnetic dipoles is proportional to the term $(3\cos^2\theta - 1)$; when θ is $54° \, 44'$, the 'magic angle', this term becomes zero, and so the dipolar broadening vanishes. For magic-angle spinning to be successful, the sample must be spun at several kilohertz—fast enough for the line between the nuclei to experience only its 'average' inclination to the field. This presents considerable technical difficulties. Nevertheless, solid-state spectra have been obtained for a variety of solids, and they show linewidths narrow enough to allow chemical shifts to be distinguished. Interpretation of the spectra is not always easy—as the chemical environment now includes interactions from adjacent molecules, and not just a single molecule, and the intensities are no longer quantitatively reliable—but it seems probable that high resolution solid-state NMR will become a very important technique in the future.

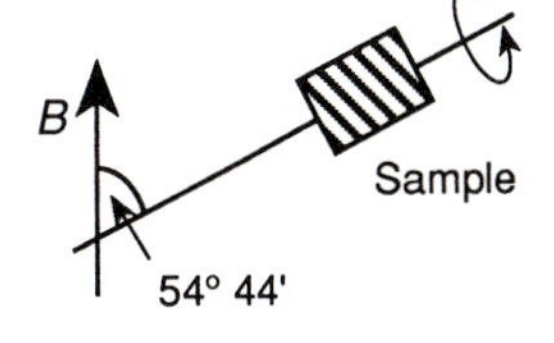

Fig. 4.14 Magic angle spinning.

4.8 Relaxation times

The equilibrium populations of the various energy levels encountered in spectroscopy are given by the Boltzmann distribution law, where the population decreases exponentially as the energy increases. In most forms of spectroscopy it is safe to assume that if these populations are temporarily disturbed in some way, for example, by the absorption of light, then equilibrium will be restored fairly rapidly. In NMR this is frequently not the case, and after the nuclei in a sample are reorientated by a radiofrequency field, it can take many seconds for the equilibrium populations to be restored. The processes by which the magnetization returns to equilibrium, or 'relaxes', have been carefully studied, partly because they have important experimental consequences, and also because they can give information on molecular motions.

For relaxation to take place, that is, for the nuclei to reorientate themselves, they must experience a suitable magnetic field. Several different relaxation mechanisms have been identified: the motions of neighbouring magnetic nuclei can cause relaxation, and this is often the dominant mechanism in organic molecules. Paramagnetic species also cause very efficient relaxation, and organic radicals and transition-metal ions are sometimes used for this purpose. Nuclei with spin $I > \frac{1}{2}$ have quadrupole moments, and these can induce relaxation. In heavy atoms, the anisotropy of the chemical shift produces fluctuating fields which can cause relaxation.

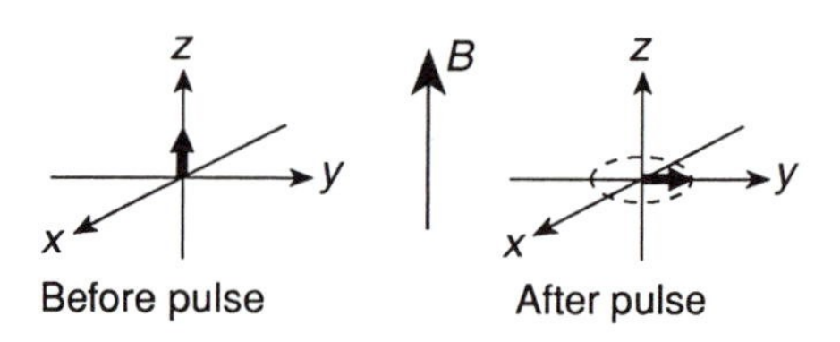

Fig. 4.15 The effect of a 90° pulse on magnetization.

The rate at which relaxation occurs is described by two relaxation times, T_1 and T_2. T_1 is called the spin–lattice relaxation time, and describes the rate at which the magnetization parallel to the magnetic field returns to its equilibrium value after a perturbation. T_2 is called the spin–spin relaxation time and describes the decay of the magnetization in the plane perpendicular to the field. Figure 4.15 shows the total magnetization of a sample in a magnetic field, with the field pointing in the z direction. Slightly more nuclei are oriented with the magnetic field than against it, and so the magnetization has the value M in the z-direction and zero in the x and y directions. A rapid radiofrequency pulse is now applied, such that the magnetization is tipped through 90°, and the magnetization now points along the y axis. The effect of the magnetic field is to make the magnetization vector rotate in the xy plane at the Larmor frequency. The magnetization now slowly returns to its equilibrium value, so M_z returns to M from zero and M_x and M_y return to zero. T_1 describes the rate of the return of M_z to M and T_2 the rate of the return of M_x and M_y to zero; T_1 is greater than or equal to T_2.

Any relaxation processes which affect the value of M_z (T_1 processes) alter the total energy of the nuclear spin system, as M_z interacts with the applied magnetic field. These processes must therefore involve exchange of energy between the nuclear-spin system and the rest of the sample, the 'lattice'. On the other hand, processes which exchange energy only between pairs of nuclear spins (T_2 processes) do not affect M_z—as the total energy of the spin system remains unchanged—but can reduce the values of M_x and M_y by causing the rotations of individual nuclei to become out of phase, and therefore to cancel each other. For mobile liquids it is often found that $T_2 \approx T_1$, whereas for solids T_2 is typically much less than T_1.

The value of T_2 limits the lifetime of an individual nucleus in its energy state, and therefore, by the uncertainty principle, affects the linewidths of NMR lines. The linewidth at half-height is $(1/T_2 \times \pi)$ Hertz; for mobile liquids T_2 may be of the order of seconds, producing line broadening of less than 1 Hz, whereas for solids T_2 may be less than 10^{-3} s, giving broad bands whose width is measured in kilohertz.

The value of T_1 is of experimental importance as it determines the extent of 'signal saturation'. We have already seen that in NMR there is a very small population difference between the different nuclear spin states; when a radiofrequency field is applied at the Larmor frequency, there is net absorption of energy, which tends to equalize these populations. At the same time, T_1 relaxation processes tend to restore the population difference. If the rate of absorption of energy is high and relaxation is slow, then the population difference decreases rapidly and the energy absorption decreases; the signal is said to be saturated. On the other hand, if the rate of absorption is lower and the relaxation is faster, then the population difference remains constant, as does the signal intensity. Samples with high values of T_1 therefore require the use of low radiofrequency fields, which can produce low sensitivity, whereas samples with low T_1 values can be studied at high radiofrequency power without saturation occurring. In some experiments small quantities of paramagnetic substances are deliberately added to the sample to reduce the value of T_1 and so avoid signal saturation.

A knowledge of relaxation times is important for the measurement and interpretation of NMR spectra. As the understanding of relaxation mechanisms has increased, measurements of relaxation times have become increasingly important. In favourable cases, the values of T_1 and T_2, and the ways in which they vary with parameters such as magnetic field and temperature, can be used to obtain information on the motions of both small and large molecules.

4.9 Basic principles of ESR

The electron spin resonance experiment involves the reorientation of the magnetic moment of an electron in a strong field; in many ways it resembles NMR, but it differs from NMR in two important details. Firstly, the Pauli exclusion principle requires that whenever two electrons occupy one orbital their spins should be opposed. It is therefore only possible to reorientate the spin of an electron if the electron is unpaired, and so ESR is restricted to ions and molecules with odd electrons. Closed-shell molecules give no ESR signal. Secondly, because the electron is much lighter than any nucleus, its magnetic moment is much greater, with the result that ESR transitions fall in the microwave region of the spectrum. The energy of an electron in a magnetic field is shown in Fig. 4.16; note that the ordering of the states $M_s = +\frac{1}{2}$ and $-\frac{1}{2}$ is the opposite of that for the proton (Fig. 4.3) as the electron and proton have opposite charges.

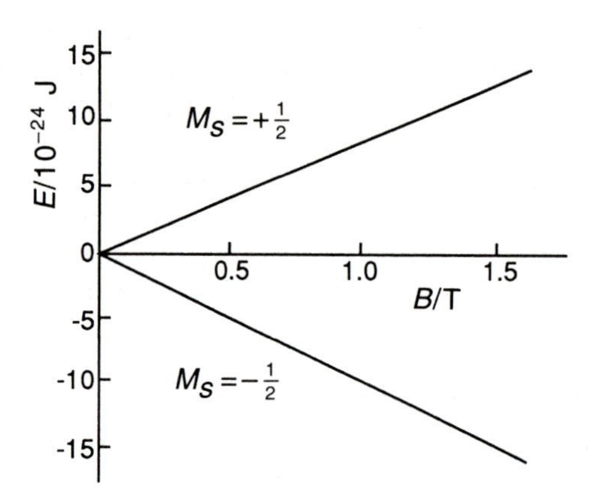

Fig. 4.16 Energies of the two states of an electron in a magnetic field.

By analogy with NMR we can write

$$h\nu = g\ \mu_B\ B,$$

where μ_B is called the Bohr magneton. The value of μ_B is given by

$$\mu_B = \frac{eh}{4\pi m_e}.$$

μ_B has the value of $9.3 \times 10^{-24}\ m^2$ A, which is 1837 times greater than μ_N, the nuclear magneton. For a free electron, g has the value 2.0023, so in a field of 1 T,

$$\begin{aligned} \nu &= \frac{2.0023 \times (9.3 \times 10^{-24}) \times 1}{6.626 \times 10^{-34}}\ \text{Hz} \\ &= 2.8 \times 10^{10}\ \text{Hz} \\ &= 28\ \text{GHz}. \end{aligned}$$

The value of g in molecules may show some deviation from the free electron value.

ESR is inherently more sensitive than NMR as the population difference between the spin states is greater. From the Boltzmann law, for an electron resonating at 28 GHz,

$$\begin{aligned} n_+/n_- &= \exp\frac{-h\nu}{kT} \\ &= \exp - \frac{(6.63 \times 10^{-34}) \times (2.8 \times 10^{10})}{(1.38 \times 10^{-23} \times 300)} \\ &= 0.9955. \end{aligned}$$

Although the populations of the two levels are still almost exactly equal, the difference between them, on which the net absorption of radiation depends, is much greater than in NMR (cf. Section 4.2). ESR spectra can be obtained readily from dilute solutions of radicals; in the most favourable cases, as few as 10^{11} spins can be detected. The probability of a transition is independent of the precise chemical nature of a radical, and so ESR can be used quantitatively to measure the number of radicals in a sample. This is of importance in the assessment of radiation damage, and in monitoring free radical reactions.

4.10 Hyperfine structure

The ESR spectra of many radicals show hyperfine structure, which is caused by the presence of magnetic nuclei in the radical; this structure is often the most important feature of the ESR spectrum. In some ways the hyperfine structure resembles the spin–spin coupling which we saw in NMR, but there are also some important differences.

In liquid samples, where there is rapid molecular tumbling, the coupling between the unpaired electron and a magnetic nucleus is due to the Fermi or contact interaction. The magnitude of this interaction depends on the value of the electron wave function at the nucleus in question. As p, d, and f orbitals all have nodes at the nucleus on which they are centred, they have zero electron density at the nucleus. They make no contribution to the electron–nuclear coupling. However, s orbitals have a finite electron density at the nucleus and electrons in orbitals with significant s character interact strongly with the nuclear magnetic moment.

If an electron interacts with just one nucleus of spin $\frac{1}{2}$, then its ESR signal consists of two peaks of equal intensity, corresponding to the two possible orientations of the nucleus. The separation of the peaks is the coupling constant A, whose value is typically measured in MHz and which is field-independent. An electron which interacts with two spin $\frac{1}{2}$ nuclei gives an ESR signal with three peaks, with intensities in the ratio of 1:2:1; these correspond to the nuclear orientations $\uparrow\uparrow$, $\uparrow\downarrow$ and $\downarrow\uparrow$, and $\downarrow\downarrow$. Figure 4.17 shows the ESR spectrum of the CH_3 radical; the electron interacts with three equivalent protons and so four peaks are seen, with intensities in the ratio of 1:3:3:1. If an electron is coupled to a nucleus with spin 1, such as ^{14}N, then the ESR spectrum consists

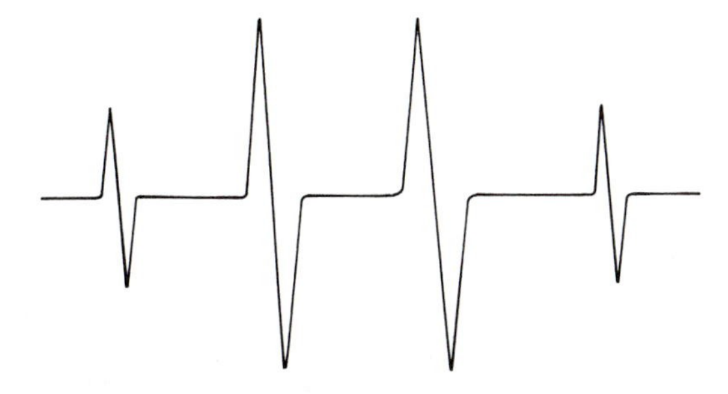

Fig. 4.17 The CH_3 spectrum.

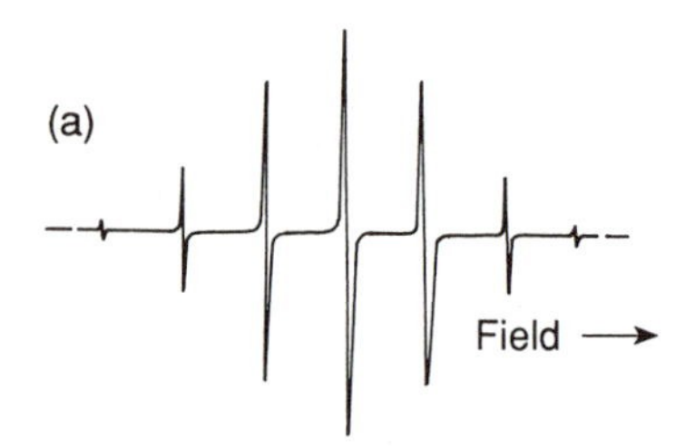

Fig. 4.18 The ESR spectrum of the benzene radical anion.

of three lines of equal intensity, corresponding to the nuclear orientations $M_I = -1, 0$ and $+1$. In many cases an electron couples to more than one set of nuclei, in which case the coupling energies are simply additive. The splitting pattern is obtained by considering the splitting caused by one set of nuclei and then superimposing on each line the splitting pattern from the second set, and so on. Complex splitting patterns can sometimes be clarified by using deuterium substitution; this produces a different splitting pattern—as $I = 1$ for the ^2H nucleus—and also alters the coupling constants, as the magnetic moment of ^2H differs from that of ^1H.

It might be thought that in aromatic radicals, where one unpaired electron occupies a delocalized π orbital, that no hyperfine structure would be observable (see Fig. 4.18). However, the π electron is found to be influenced by the magnetic moment of the protons in the ring by the mechanisms of 'spin polarization'. Figure 4.19 shows the spin arrangements in a C–H fragment in an

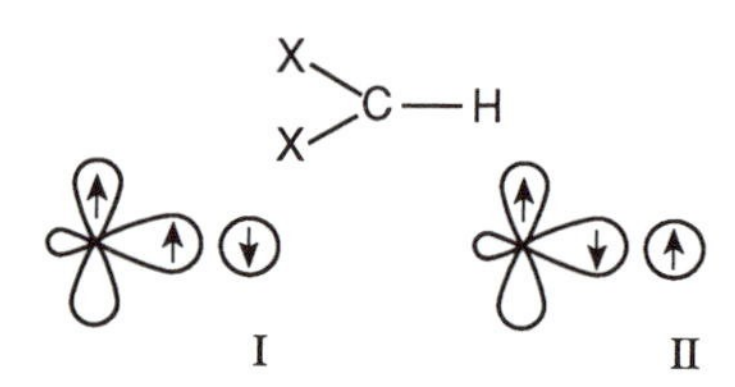

Fig. 4.19 The origin of negative hyperfine coupling in C–H fragments: I is favoured over II.

aromatic molecule. The presence of an unpaired electron in the p orbital on the carbon atom causes the C–H bond to be spin polarized, with the electron of one spin marginally closer to the C nucleus and the electron of the other spin closer to the H nucleus. The electron closer to the C atom has the same spin as that in the p orbital, as this arrangement has a more favourable exchange energy. The electron closer to the H nucleus, therefore, has the opposite spin and this electron interacts with the proton magnetic moment. Thus in the radical ion $C_6H_6^-$, the unpaired electron interacts with six equivalent protons and gives a spectrum consisting of seven lines.

One of the most important applications of ESR is the identification of the chemical structures of radicals. The hyperfine splitting can be of great value here as it contains information on the types and numbers of magnetic nuclei and on their environments. ESR also gives information on the electronic structures of radicals. The value of the coupling constant for a free hydrogen atom is 1420 MHz; if the value of the coupling constant of an electron to a hydrogen atom in a radical is A MHz, then we may crudely say that the spin population on that hydrogen atom is $(A/1420)$, or that the coefficient of the 1s orbital based on the atom is the molecular orbital occupied by the unpaired electron and is $(A/1420)^{\frac{1}{2}}$. Similarly, the coupling constant between a 2s electron and a ^{13}C nucleus is 3330 MHz; an electron in an sp^3 hybrid orbital would give a value of one-quarter of this, as the orbital has 25 per cent s-character. Coupling constants therefore give information on the hybridization in a radical, and hence on molecular geometry. It is also possible to estimate the degree of covalency in complex ions by observing the coupling of an unpaired electron to a ligand nucleus.

One of the best known applications of coupling constants is the use of the McConnell equation for calculating spin populations in aromatic systems. McConnell suggested that the coupling constant A_{H} between an unpaired electron and a proton in an aromatic ring was simply proportional to the spin population of the p orbital on the adjacent carbon atom, p_{C}. Thus

$$A_{\mathrm{H}} = Q p_{\mathrm{C}},$$

where Q is roughly constant for all aromatic radicals. The value of Q can be shown to be -63 MHz by considering benzene, in which the spin population on each carbon atom is one-sixth by symmetry. Spin populations measured in this way may be used to test theoretical methods such as Hückel calculations; agreement in many cases is good.

In solid samples there is a further direct dipolar interaction between the magnetic moments of an electron and nucleus, which is not now averaged to zero by molecular tumbling. However, the magnitude of this interaction is often relatively small compared with the Fermi term, and for a sample where the radicals are not oriented, the overall effect on the spectrum is to produce broader lines. If the radicals are oriented within a crystal—which can occur if a pure solid is subjected to radiation damage—then sharper lines are obtained, and the spectrum changes as the crystal orientation is altered. Details information can now be obtained on the structure and environment of radicals.

4.11 The g value

In NMR we saw that a nucleus in a molecule does not resonate at exactly the same frequency as in a free atom; this difference is the chemical shift and is treated by regarding the field experienced by the nucleus as being different from the external field. In ESR a similar phenomenon is observed, but it is more usual to treat deviations from the free electron case as a property of the electron and not the field. Therefore we write

$$h\nu = g\,\mu_{\mathrm{B}}\,B,$$

where g measures the ratio of the resonant frequency to the applied field. If the value of g deviates from the free-electron value, then the position of the ESR signal in the spectrum shifts to high or low frequency, in a way analogous to the chemical shift in NMR.

The g value of an electron will differ from the free electron value if the applied magnetic field induces orbital motions which produce an extra magnetic field with which the electron spin will interact. In many large organic radicals and first-row transition-metal ions this orbital contribution is very small and the g values differ little from the free electron value; thus for NO_2, $g = 1.999$, and for CH_3, $g = 2.00255$. Differences in g values in these cases produce relatively small shifts in ESR spectra; for example, the difference in resonant frequency at 1 T between CH_3 and NO_2 is about 50 MHz, whereas the total width of the CH_3 hyperfine structure is about 130 MHz. Determination of g values can be of use in identifying free radicals, in the same way as chemical shifts are used in NMR, but in many cases the hyperfine structure is of greater value as it contains more detailed information.

The g values for the later transition-metal ions, and for the lanthanides and actinides, show much greater deviations from the free electron value. The situation is further complicated if there is more than one unpaired electron in an ion or if the spin–orbit coupling is very strong. The g value is often strongly anisotropic, and if a solid sample is used, its value in the x, y, and z directions may be determined separately. ESR can give very detailed information on the electronic structure of ions with partially filled d and f orbitals, but the interpretation is now complex and each configuration has to be considered separately.

Both NMR and ESR have received tremendous attention in recent years from chemists, physicists, and biologists alike. They are topics which require a great deal of study for a full understanding of their applications. Although this chapter is only an introduction, further details can be found in more advanced texts.

Further reading

P. W. Atkins, *Quanta: a handbook of concepts*. Oxford University Press (1991).

F. Hund, *The history of quantum theory*. Harrap & Co., London (1974).

G. C. Schatz and M. A. Ratner, *Quantum mechanics in chemistry*. Ellis Horwood/Prentice Hall, Hemel Hempstead (1993).

T. P. Softley, *Atomic spectra* (Oxford Chemistry Primer Series). Oxford University Press (1994).

M. J. Winter, *Chemical bonding* (Oxford Chemistry Primer Series). Oxford University Press (1993).

C. A. Coulson, *The shape and structure of molecules* (revised by R. McWeeny). Oxford University Press (1982).

J. M. Hollas, *Modern spectroscopy*. Wiley, New York (1991).

G. Herzberg, *Infrared and Raman spectra of polyatomic molecules*. Van Nostrand, New York (1945).

E. B. Wilson, J. C. Decius, and P. C. Cross, *Molecular vibrations*. McGraw–Hill, New York (1955).

G. Herzberg, *Spectra of diatomic molecules*. Van Nostrand, New York (1950).

G. Herzberg, *Electronic spectra and electronic structure of polyatomic molecules*. Van Nostrand, New York (1966).

A. G. Gaydon, *Dissociation energies*. Chapman and Hall, London (1968).

R. J. Abraham, J. Fisher, and P. Lofthus, *Introduction to NMR spectroscopy*. Wiley, New York (1991).

R. K. Harris, *Nuclear magnetic resonance spectroscopy*. Longman, London (1986).

A. E. Derome, *Modern NMR techniques for chemistry research*. Pergamon, Oxford (1987).

Index

There are separate sections in this index for *energy levels* and *quantum numbers*.

Energy levels in:

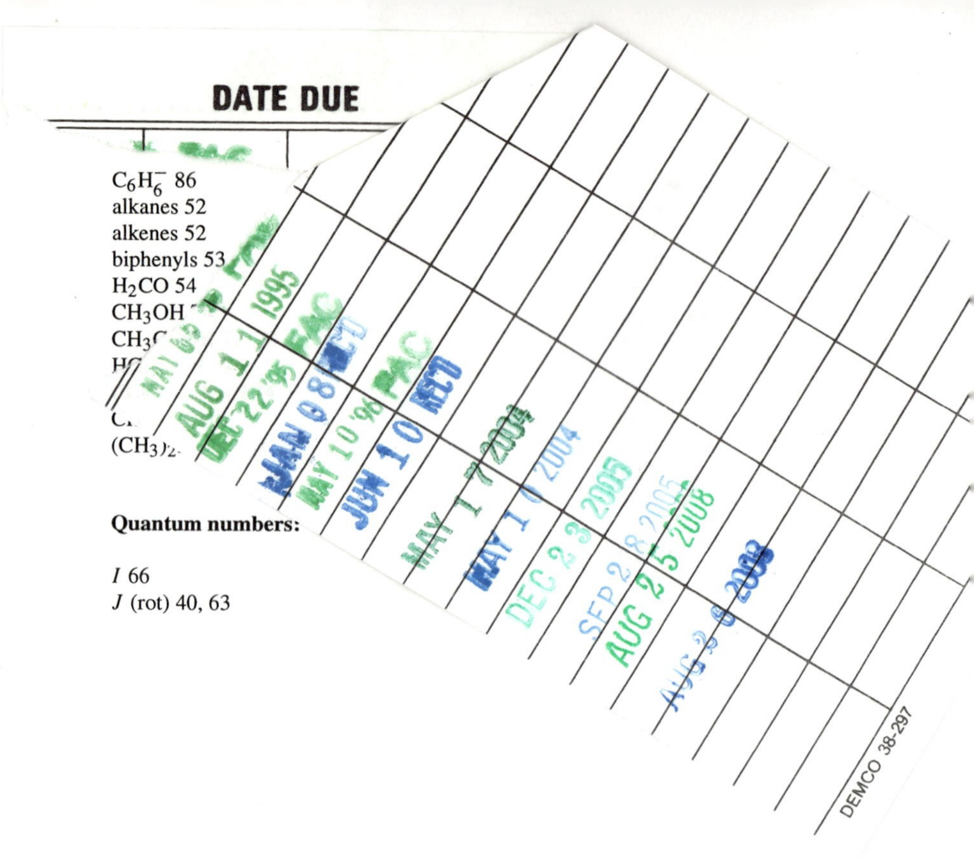
DATE DUE
DEMCO 38-297

BGSU Libraries
A11321777325